RECENT ADVANCES IN AUTISM SPECTRUM DISORDERS - VOLUME II

Edited by **Michael Fitzgerald**

Recent Advances in Autism Spectrum Disorders - Volume II
http://dx.doi.org/10.5772/50854
Edited by Michael Fitzgerald

Contributors

Katarina Dodig-Ćurković, Mario Ćurković, Josipa Radić, Marianna Boso, Enzo Emanuele, Elizabeth Barron, Noemi Piaggi, Giulia Scanferla, Matteo Rocchetti, Umberto Provenzani, Davide Broglia, Paolo Orsi, Roberto Colombo, Sara Pesenti, Marta De Giuli, Elena Croci, Stefania Ucelli, Pierluigi Politi, Jenny Secker, Francesco Barale, Jane Yip, Fengyi Kuo, Betsy Powers, Mark Tommerdahl, Eric Francisco, Penny Spikins, Ian Richard Freckelton, Donald Rojas, Kristina McFadden, Segado Segado Vasquez, Shunit Reiter, Nirit Karni

Published by InTech
Janeza Trdine 9, 51000 Rijeka, Croatia

Publishing Process Manager Viktorija Zgela
Technical Editor InTech DTP team
Cover InTech Design team

Additional hard copies can be obtained from orders@intechopen.com

Recent Advances in Autism Spectrum Disorders - Volume II, Edited by Michael Fitzgerald
p. cm.
ISBN 978-953-51-1022-4

For EU product safety concerns:
IN TECH d.o.o., Prolaz Marije Krucifikse Kozulić 3, 51000 Rijeka, Croatia,
info@intechopen.com or visit our website at intechopen.com.

Meet the editor

Michael Fitzgerald was the first Professor of Child Psychiatry in Ireland at Trinity College Dublin. He trained at Mount Sinai Hospital, Chicago and Maudsley Hospital, London. He continues to practice in Child and Adult Psychiatry and carries out research on autism and it's co-morbidities. He has diagnosed over 2,500 persons with ASD in the past 40 years. He is on the editorial board of a number of journals and has published extensively in peer-reviewed journals as well as written, edited or co-edited 20 books.

Contents

Preface

There is no doubt that the technological and scientific progress of homosapiens has been hastened by persons with Autism Spectrum Disorders (Fitzgerald 2000 (a & b) 2004, 2005). Spikins in her chapter illustrates the Stone Age origins of autism with reference to the archaeological record and the Palaeolithic record. It is of critical importance to pay attention to the positive aspects of autism.

The study of attitudes of people towards autism is of critical importance as the issue of a stigma is still very prevalent in all countries even more so in non-western countries. Karni-VizerNirit et al describes the attitudes towards autism among Israeli Arab teachers of college students. They are particularly focused on inclusion. Interest in religious faith gave some mothers strength. Coulthard and myself demonstrated this also some years ago (Coulthard and Fitzgerald 1979, 1999). This study was also done on mothers of children with Autism Spectrum Disorders.

Autism Spectrum Disorder is a brain based disorder. Francisco et al discusses the critical importance of cortical modularity and information processing. They emphasise the smaller than normal mini columns and deficient cerebral cortical GABAergic inhibitory neurotransmission. The issue of the mini columns also came up in a paper by Casanova et al 2007 in relation to creativity. McFadden et al have written a complimentary chapter on the electrophysiology of autism. Indeed they point out that EEC studies in the future may help clinicians to separate high-risk infants from typically developing infants. In terms of medical treatment Dodig-Curkovic et al provide a comprehensive overview of medical treatment and autism. It is often necessary to use medications in persons with autism.

Problems in toilet training are very common in autistic children. Yip et al explored this from the sensory experience point of view in persons with Autism Spectrum Disorders. They propose a putative sensory-neural understanding of the toilet training resistant cohort. Boso et al describes the recovery orientation of a farm community for persons with severe autism. Farm communities have considerable history for example in Ireland there was one developed by the Irish Society for Autism and Pat Matthews. These communities provide impressive quality of life for persons with autism. Boso et al also described data using the Developing Recovery Enhancing Environment Measures-Italian Version.

It is well know that persons with ASD get into difficulties with the law (Fitzgerald 2010) and that there are higher rates of persons with ASD in prison than in the community. Freckelton provides a very useful chapter from the court perspective of the forensic issues in relation to Autism Spectrum Disorders. This gives a unique insight into a complex topic.

There are a relative paucity of papers on architecture on autism. Persons with autism are very sensitive to the built environment and sensory issues are particularly relevant here. The chapter by Segado Vazquez makes particularly interesting reading.

There is an overlap between autism and schizophrenia (Fitzgerald 2012) - the view that they were separate is incorrect. It is disappointing that there is no biological marker for ASD and that genetics has no influence on the treatment of ASD which is an extremely complex disorder.

Genes have multiple affects and often can give the features of ASD and creativity at the same time. It is important that the positive as well as the disability elements are discussed with families.

While the vast majority of persons with autism are highly moral and ethical, a tiny group can get involved in serious crime e.g. school shootings, multiple murders etc. (Fitzgerald 2003, 2005 (a), 2005 (b), 2010). It is critical to take this group seriously and to take their threats seriously to avoid a public catastrophe as we have seen in many countries.

Professor Michael Fitzgerald
Consultant Child & Adult Psychiatrist
Trinity College, Dublin
Ireland

Section 1

Evolution

Chapter 1

The Stone Age Origins of Autism

Penny Spikins

Additional information is available at the end of the chapter

http://dx.doi.org/10.5772/53883

'Their strengths and deficits do not deny them humanity but, rather, shape their humanity'

Grinker 2010: 173 [in [1]]

1. Introduction

1.1. Minds from a stone age past

Our modern societies have been said to house 'stone age minds' (see [2]). That is to say that despite all the influences of modern culture our hard wired neurological make-up, instinctive responses and emotional capacities evolved in the vast depths of time which make up our evolutionary past. Much of what makes us 'human' thus rests on the nature of societies in the depths of prehistory thousands or even millions of years ago.

Looking back on the archaeological record of the early stone age there is much to be proud of in our ancestry. Not only our remarkable intelligence but also our deep capacities to care about others and work together for a common good come from evolutionary selection on early humans throughout millions of years of the stone age. As far back as 1.6 million years ago we have archaeological evidence from survival of illnesses and trauma that those who were ill were looked after by others, and by the time of Neanderthals extensive care of the ill, infirm and elderly was common, see [3,4]. From at least one million years ago we see evidence for widespread collaboration in hunting, in sharing food and in looking after increasingly vulnerable young. Stone age societies, much as recent hunter-gatherers such as the Selk'nam of Tierra del Fuego (figure 1), lived in small groups who cared deeply about each other, and worked together to survive.

Nonetheless our evolved minds also have a darker side. For most of our early existence, at least until only around 100,000 years ago, human groups were relatively isolated, and much of our common drive to identify 'us' and 'them' probably has its roots in a suspicion of 'others' which dates to this time. Studies of the remains of a group of *Homo antecessor* dating to around 900,000 years ago at Atapuerca in northern Spain for example have revealed that these people probably hunted and ate neighbouring groups to defend their territories (see [5]). Small wonder that as a result we find ourselves far too often being afraid of those who we feel are different from ourselves. Our neurological response to the pain of others for example can be tempered by whether we see them as belonging to the same group as ourselves or not (see [6]) and if we see people as different to ourselves we can even feel a sense of pleasure at their pain (see [7]).

Thanks to our capacities for self-awareness and moral judgement we can make balanced decisions about how we treat others. Undoubtedly we must also influenced by our more recent evolutionary history of a remarkably widespread collaboration. Indeed from 100,000 years ago onwards we begin to see evidence for widespread links across different stone age groups in many different parts of the world. In ice age Europe for example around 35-10,000 years ago marine shells travel over 2000km through exchange networks which helped provide a social buffer to withstand shortfalls in resources (see [8]). Somehow these groups overcame their tendencies to distrust outsiders and worked out ways of working together.

http://dx.doi.org/10.5772/53883

Figure 1. Selk'nam hunter-gatherers from Tierra de Fuego on the move. Hunter-gatherer societies such as the Selk'nam depend on high levels of collaboration for their survival. But do all minds need to be the same for collaboration to work, or are different minds a better recipe for success?

Our remarkable abilities to extend ourselves to care about others' wellbeing can sometimes be rather fragile. As the same time as being able to care about global issues or the wellbeing of those we have never met, we can work hard to set up divisions which set us above others. One can't help but wonder if future societies may well look in disbelief at the plethora of ways in which the twentieth and twenty-first centuries have found more detailed and elaborate ways to define a mentally 'normal' mind in contrast with a mentally different (and by implication 'wrong') other. Whilst our abilities to deal with mental health issues have become ever more sophisticated, our 'labelling' of many conditions as disorders can fly in the face of the more obvious reality that the human condition involves a great deal of suffering, and not all of that suffering can be seen as 'unnatural'.

Many so-called 'disorders' may be a natural part of humanity. Conditions such as anxiety or depression are unwelcome but far from unnatural for example. Equally, genetically linked conditions such as schizophrenia or bipolar disorder appear to have a long history, with good evidence to suggest a role for those less grounded in reality in hunter-gatherer societies as shaman (see [9]). Though a shaman's apparent difference, and connection to another world, may give them power, the trances experienced by shaman, and seen as providing a link to the spirit world are more commonly painful than pleasurable. Their experience and behaviour may have at times given them a certain social role in the past, but the same experience is more typically seen as a disorder today.

It is within the context of a fashionable drive to label and classify 'disorder' that the label of autism has emerged. Yet are autistic minds really 'abnormal' or 'wrong'? We would be well advised to be cautious of media warnings of an autism 'epidemic', wording which easily conjures up a picture of a growing disease threatening society. There is every reason to suggest in contrast that what makes 'us' human is not a single 'normal' mind but a complex interdependency between different minds in which autism plays a key role (see [10, 11, 12]). As Grinker (see [1]) illustrates autism should not deprive someone of humanity, but rather shape their (and our) humanity. Whilst the 'story' of autism is nearly always told as beginning with its labelling in diagnosis by Kanner and Asperger in the early 19th century (see [13]) autism may have much older roots, and a more significant role to play in the emergence of our species. This much earlier 'story' of the role of autism is an important one which accords a key role to autism in the emergence of humanity.

This chapter explores the potential stone age origins of autism, considering the contribution which those with autism may have made to small scale prehistoric societies and the archaeological evidence for a long time depth to the influence of autism.

2. Autism and society

Is autism part of what makes us 'human' today? Autism has traditionally been seen as a condition of people somehow *outside* society. However recent research has challenged this view, suggesting in contrast that autism is part of the processes that allow societies work together.

Various authors have questioned whether autism, particularly high functioning autism or aspergers syndrome, should always be seen as *a disability* (see [14, 15, 11, 12]). That is not to say that life with aspergers syndrome is not often difficult or challenging or that coping with having such a condition in a social world is not often distressing but that at least *at times* it can sometimes be an advantage to have an 'exact mind' (see [16]). Aspergers syndrome often brings with it particular talents in a focus on detail, understanding of systems or abilities to concentrate on a particular problem (see [11]) and is associated with heightened awareness of details, including musical pitch as well as sensory sensitivity (see [17]). High rates of those with aspergers syndrome characterise occupations such as engineering and mathematics (see [18]) as well in universities and the legal system (see [19]). There may be many situations in which having aspergers syndrome makes life difficult, but many of those with aspergers syndrome have a place in society, making a valuable contribution.

Are people with autism motivated to be part of a greater social good? One of the key misconceptions of autism is that a 'lack of empathy' carries with it a tendency to care far less about others wellbeing than the 'neurotypical' might. However empathy comprises a cognitive and an affective component (understanding others feelings and caring about others feelings). The affective component, how much someone will care about others, has been shown to be intact in autism (see [20]). People with autistic spectrum conditions 'care' about others wellbeing as much as anyone else might (even if their abilities to intuitively sense others feelings are impaired), often channelling such concerns into wider social endeavours such as a drive for fairness and justice (see [11]) or scientific progress. Autism implies that people care about others *in a different way*.

Are people with autism really part of human social life? A further misconception is that those with asperger's syndrome are unsocial. Anthropological studies of autism have made a significant impact on our understanding of what it is to be autistic and social and shown that whilst 'autistic sociality' is notably *different* those with autism are not less social. Autistic sociality may often be focused on exchanging knowledge rather than sharing feelings or extended narratives, and is often mediated through the material world (today made up of books or computers). Autism means that people are social *in a different way* (see [13]).

Even those with severe autism can share a sociality which binds them to others. Solomon (in [21]) for example describes Sacks account of two severely autistic twins, John and Michael, who were institutionalised from childhood. In their early twenties and delighted in sharing mathematical concepts and 'conversing' in prime numbers.

Sacks writes:

They were seated in a corner together, with a mysterious secret smile on their faces, [...] enjoying the strange pleasure and peace they now seemed to have.[...] They seemed to be locked in a singular, purely numerical, converse. John would say a number—a six-figure number. Michael would catch the number, nod, smile and seem to savour it. Then he, in turn, would say another six-figure number and now it was John who received and appreciated it richly. They looked, at first, like two connoisseurs wine-tasting, sharing rare tastes, rare appreciations. (Sacks 1970: 202, in [22])

The twins happily welcome Sacks to their conversation when he joins in with his own prime numbers. They provide an example of how apparently extremely autistic individuals can connect socially to others, and derive pleasure from their social contribution and connection, albeit in a non-typical manner. Sadly the different nature of their communication lead them to be separated to prevent them communicating in this 'non-normal' manner.

Whilst it is clear that for the twins pleasurable social life may be distinctive from the norm, in many ways this type of connection is not uncomparable to motivations and pleasures of scientific endeavour as described by Nikola Tesla (see [23]).

I do not think that there is any thrill that can go through the human heart like that felt by the inventor as he sees some creation of the brain unfolding to success … Such emotions make a man forget food, sleep, friends, love, everything … I do not think you can name many great inventions made by married men. (see [24]).

Not only rare geniuses like Darwin or Einstein may have been autistic (see [25, 26, 27, 28]) but many more common and far less obviously distinctive members of society may have minds that are distinctively different from what we see as 'typical', and add something critical to what makes us 'human'.

2.1. Inherited gifts of insight and action?

'I feel sure that my way of being is only a disability of context, that what have been labelled symptoms of autism in the context of my culture are inherited gifts of insight and action' Dawn Prince (in [29])

As Dawn Prince comments, autism can be seen as a disability of context. How we view expressions of autism, whether as laudable and productive (such as an extreme focus on scientific discovery to the exclusion of other concerns for example) or as unproductive and threatening (in the case of the mathematical communication of the twins studies above) is greatly dependant on our culture. Whereas all cultures will have some limits to the nature of unusual behaviours that could be readily supported by others (and severe autism may never be an advantage), some may have been more accommodating of autistic difference than others, and in turn benefitted from what autism may have brought to society (figure 2).

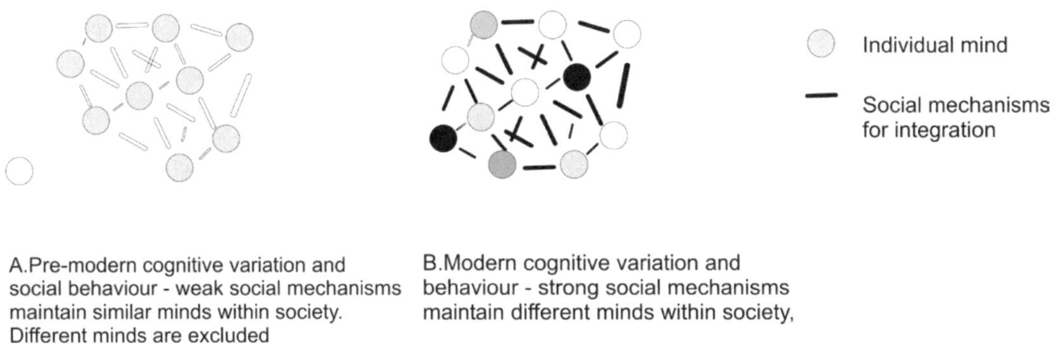

Figure 2. Social means of accommodating autism (or other differences in mind) and the resulting nature of cognitive differences.

Dawn Prince's observation that autism brings with it inherited gifts of unique *insight* and *action* itself gives us a unique insight into the potential contribution which *in certain contexts* autism may have made societies in the far distant past as well as the present.

2.1.1. Unique insight

Baron-Cohen (in [30]) describes the main areas of talents in autism. These talents derive from a drive to understand systems, and illustrate the domains in which unique insights often lie.

The major kinds of system focused on by those with autism include:

- *collectible systems* (e.g. distinguishing between types of stones or wood);
- *mechanical systems* (e.g. a video recorder or a window lock);
- *numerical systems* (e.g. a train timetable or a calendar);

http://dx.doi.org/10.5772/53883

- *abstract systems* (e.g. the syntax of a language or musical notation);
- *natural systems* (e.g. the weather patterns or tidal wave patterns);
- *social systems* (e.g. a management hierarchy or a dance routine with a dance partner);
- *motoric systems* (e.g. throwing a Frisbee or bouncing on a trampoline).

Restricted or circumscribed interests shown by children with autism (figure 3), and which tend to fall into these domains may be difficult for parents to manage, but are often also seen as related to unique strengths or talents (see [31, 32]). An obsession with weather patterns as a child for example may in some lead to a particular strength and an academic focus on meteorology as an adult. Many famous scientists appear to show autistic traits. Isaac Newton's unique insights into astronomy for example derive from a particularly focused motivation to understand the systems behind the movements of astronomical features (figure 4).

Figure 3. An eight month old boy with autism obsessively stacking cans (source: Wikimedia commons)

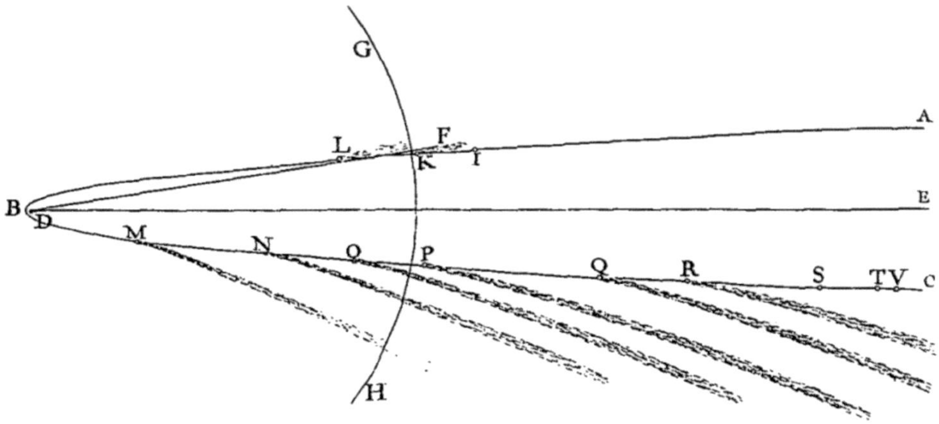

Figure 4. Sir Isaac Newton's depiction of the orbit of the Comet of 1680, fit to a parabola. *The Mathematical Principles of Natural Philosophy.* London: Benjamin Motte, 1729: 358. (Wikimedia Commons)

In a stone age context a focus on fine scale technological prowess or understanding and prediction of nature has an obvious contribution, particularly in harsh and risky high latitude environments such as the Arctic where survival depends on technology occupation depends on highly technological systems. The Inuit for example use complex multi-component harpoons to fish for seal, finely engineered dog sleds and highly efficient hunting equipment.

2.2. Autism and material culture

Though subtle, certain differences in the material culture selected and created by those with autism reflect their different minds. Thus we can argue that it ought to be possible to discriminate the material record of a society which includes or even encourages autistic traits from those where autism is unsupported.

Children and adults with autism use and create the world around them in subtly different ways, though it is children who have been studied most intensively. Children diagnosed with autistic spectrum disorder tend to engage differently with toys, for example focusing on spinning objects or lining up toys, and seem to derive comfort from *precise ordering* or regular patterning (see [33]) as seen above. Adults in turn, even if high functioning and not usually detectable as 'different', relate differently to the material world around them, tending to find comfort in ordered patterns. A focus on understanding systems leads to *detailed record keeping and scientific insights*. Baron-Cohen (in [30]) for example notes the precise recordings of weather patterns in the notebooks of Kevin Phelps and equivalent focus often drives scientific genius. A drive to understand and experiment is related to the creation of *inventions or technological innovation* (with aspergers syndrome being associated with families of engineers, see [18]).

http://dx.doi.org/10.5772/53883

Differences in perception also influence the creation and use of objects. Children with autism notice the numbers on telegraph poles, and differences in perception lead to adults *noticing and dealing with finer details* than others might. *Autistic art* is thus notably distinct. The art of Nadia, an autistic 'savant' for example is typical in being extraordinarily detailed, in common terms a representation of 'the parts' rather than 'the whole' (or the trees rather than the wood), figure 5, and in contrast to figure 6 (in [34]). The same pattern is seen in the art of Peter Myers (see [16]), who also shows remarkable talent in embedding illusions within his work. Kellman (in [35]) argues that differences in visual perception creates the distinctive features of autistic art, alongside the unique focus that a lack of perception of some other areas of external environment can bring.

Figure 5. Horse and rider completed at approximately 5 years 6 months by Nadia. Selfe 2011: figure 2.7: p32 (with kind permission Lorna Selfe).

Figure 6. Two riders drawn by non-autistic children aged 6 years. Selfe: 2011 (figure 2.10, p35). With kind permission Lorna Selfe.

2.2.1. Action

As well as driving detailed recording, fine scale understanding and new innovations of the natural world, autism may also be related to *particular types of action* in other more social ways. Whilst empathising leads to tendencies to follow allegiances (see [36]) autism leads to a focus on strict fairness in social relationships regardless of any particular allies (see [11]). An autistic creation of rigidly clear rules and obsessively fair social behaviour may thus play a key role in defining the 'rules' or legal systems which allow cooperation between unfamiliar people and constrain exploitation, explaining an association in the present between asperger's syndrome and the legal profession (see [19]). Those with autism appear to play a key role in the creation and enforcement of rigid social rules.

Clues to the significance of defined rules of social behaviour for stone age societies can be found in modern hunter-gatherers. Amongst the Inuit for example, as with most hunter-gatherers, connections to external groups and collaborations at times of crisis work through rigid systems of defined behaviour, rather than being driven by a far messier suite of complex allegiances or personal favours.

http://dx.doi.org/10.5772/53883

In the northern Canada the Netsilik Inuit for example had a highly rule based system called *niqaiturasuaktut* which is used to ensure 'fair' sharing during collaborative winter seal hunts (see [37]). After a hunter kills a seal sharing partners are defined by a combination of inheritance and naming and decided by the males of the family. A particular woman divides the carcass and actually shares it with the other partners. The division must follow specific rules. The seal meat and blubber is divided amongst 14 partners, with the first 7 being the most important. The hunter himself only keeps the flippers, so relying on a repeat of the system in future hunts to provide him with meat for himself. These elaborate and rigidly defined rules prevent emotionally driven personal allegiances from influencing the sharing of resources and provide a system by which those who might not usually work together can collaborate for a common good.

The incorporation of an autistic obsession with fairness and rules may have been key to providing highly systemised conventions to circumvent any tendencies to follow allegiances, or react emotionally to the unfamiliar and so may have been critical in promoting collaboration between different groups. It is not unreasonable to suggest that the widespread connections, exchange of materials and collaboration at times of need which we see after around 100,000 years ago may have been driven by the inclusion of autistic minds into societies. These types of systems may have been the key for example to allowing upper Palaeolithic groups to collaborate across large regions to survive local famines during the severe environments of ice age Heinrich events (see [38]).

Autism may often be a disadvantage where intuitive understanding of others is important, and can be unhelpful to the emotional wellbeing of others. Nonetheless where they could be integrated and supported, at a certain level a few individuals at the extreme of the spectrum of mind may have made an important contribution to past societies both in technological and social domains.

3. Autism and the archaeological record of the Palaeolithic

In the light of the potential value of autistic insight and action in certain contexts it is possible to view the archaeological record rather differently. Rather than a progressive sophistication of a single human 'mind', a more plausible explanation for much of the patterning in the archaeological record is as the marked emergence of autistic traits within a modern 'humanity' made up of complex interrelationship between *different minds*.

The earliest evidence for any autistic characteristics emerges well after the split between our own species and our nearest relatives the Neanderthals (occurring around 500,000 years ago), perhaps unsurprisingly as some of the key genes for autism have been found to be lacking in the Neanderthal genome (see [39]) and that of the other closely related species to modern humans, the Denisovans (see [40]).

However after 100,000 years ago various elements of the archaeological record document certain new traits which appear to be linked to autism – such as a unique focus on detail, technological innovation, and understanding of complex systems (see [10]) as well as evidence for large scale collaborations in the exchange of materials between groups (see [41]). Many of these new elements can be associated with what has been termed the appearance of 'modern human behaviour'.

After around 100,000 years ago we begin to relatively suddenly see the emergence of 'inventions' such as the spear thrower, multi-component harpoon and tiny microlithic stone tools (figure 7 & 8) which appear to have been essential for the colonisation of previously unoccupied regions such as the far north (see [42]).

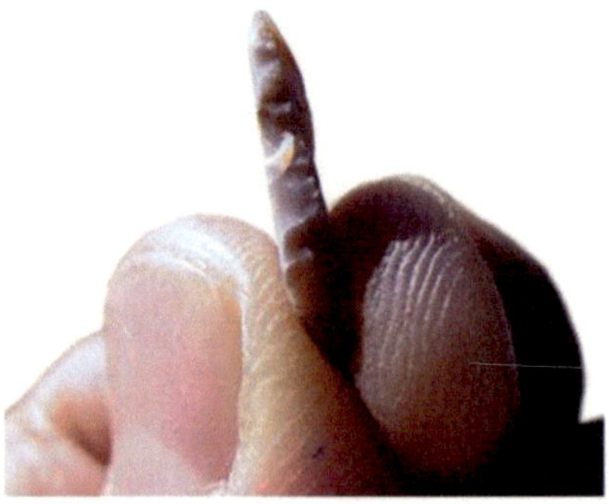

Figure 7. A 'microlith', these tiny tools formed part of barbs in arrow shafts, as well as other uses, and were highly efficient ways of making effective hunting weapons as well as maximising the use of stone tools materials and the efficiency and maintainability of tools with individual microliths being replaceable. *With kind permission José-Manuel Benito Álvarez.*

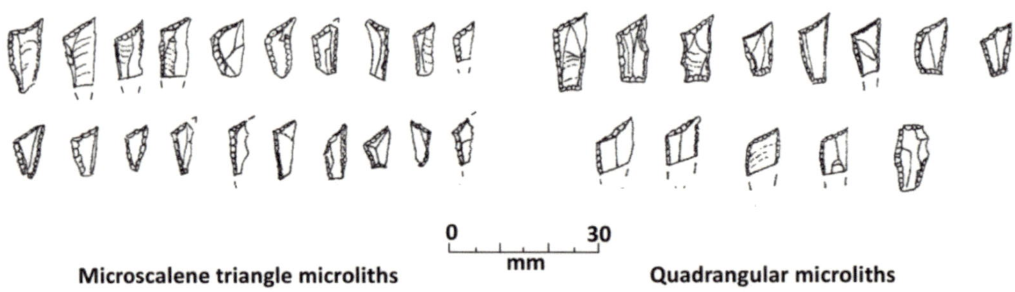

Figure 8. Microliths, forming part of highly engineering technologies, only appear after about 100,000 years ago. These microliths are from Red Ratcher Late Mesolithic site in the Pennines (courtesy of Paul Preston).

For example changes in 'modern human' technology include the appearance of tiny microlithic points such as at Howieson's Poort and Rose Cottage in South Africa at around 75,000 years ago. Other innovations in Africa at his time including finely made bifacial points made

http://dx.doi.org/10.5772/53883

on raw materials which may have been derived from structured exchange networks, bone tools, new symbolic art such as engraved patterns on ochre and ostrich eggshell and the formal ordering of space on sites (see [44]). Microliths and other elements of 'modern' behaviour are also later found at Patne in India, following the presumed colonisation of the southern corridor by fully modern humans (see [45]).

Somewhat later in Ice Age Europe, around 35-10,000 years ago, we see potential further evidence for a stone age context in which autism may have played an important social role.

Firstly, a number of artefacts found at European Upper Palaeolithic sites illustrate a unique focus on recording and understanding natural systems, particularly astronomical systems, which parallels with that seen in those with aspergers syndrome today. The Taï plaque for example, a 9cm long engraved bone from the Grotte de Taï, dated to around 10,000 years ago (see [43, 46]) is covered with many notches interpreted as a calendrical notation spanning over a year (figure 9).

Figure 9. The Taï plaque (Marshack 1991: figure 1, p26.[43])

The Abri Blanchard plaquette, dated from around 32,000 years ago is perhaps even more remarkable. The patterns on this bone record the phases of the moon and its position in the sky related to a notched co-ordinate system at the edge of the plaque (figure 10, see [47, 46, 48]).

As well as other artefacts which also appear to carry calendrical, astronomical or other notation there are also other hints of autistic influence. The Raymonden plaquette from around 12,000 bp for example illustrates an autistic like approach to social relationships. This bone features an extended bison skeleton, with figures sitting on either side of the spine, illustrating both a detailed anatomical knowledge of anatomy (showing individual vertebrae) and with a focus on equal or systematically defined sharing (figure 11).

The most famous example of a link between autism and the contemporary archaeological record of this period however comes from the famous art. Upper Palaeolithic art in south-western europe is dominated by often extraordinarily realistic and naturalistic depictions of animals, both on cave walls (see front figure and figure 12) and in portable art (see figure 13). A number of elements of this art, such as highly realistic detailed figurative representation, a focus on parts (with drawings often overlapping) and a remarkable visual memory from what can only have been limited opportunities to note details of dynamic animals are found in common with autism (see [49, 35, 50, 10]). Whilst we might not necessarily suggest that the ice age artists were autistic themselves, it would not be unreasonable to conclude that *autistic perception* and the influence of those with autism on society had a significant influence on the style of art.

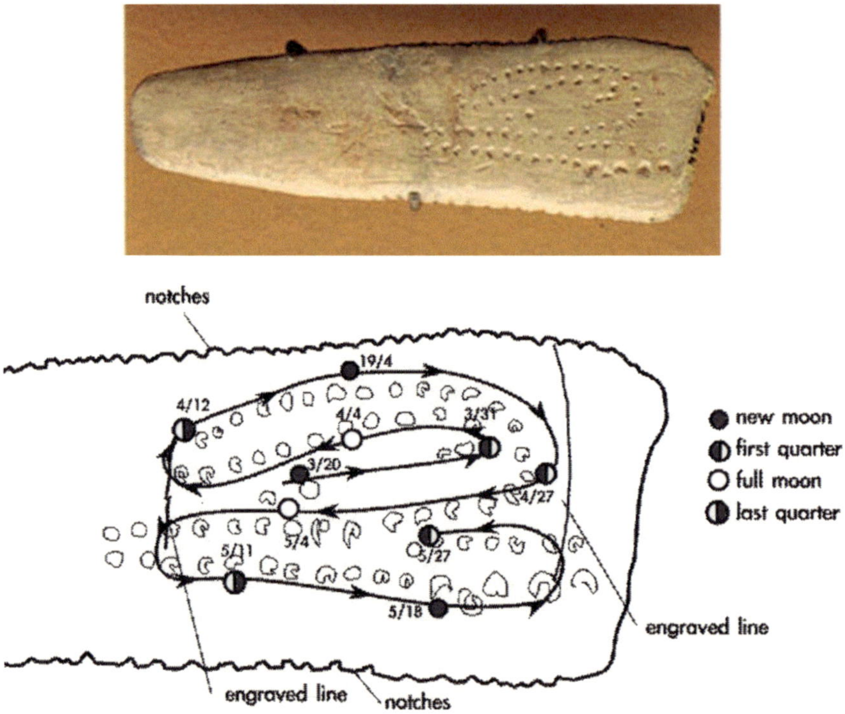

Figure 10. The Abri Blanchard plaquette (De Smedt and Cruz 2011: figure 1, with kind permission).

Figure 11. Raymonden plaque, c10,000 years old, Raymonden, Dordogne, southern France (image: museo de Altamira).

http://dx.doi.org/10.5772/53883

Figure 12. B. & G. Delluc viewing frieze of swimming reindeer in Lascaux cave, Dorgdogne c 20,000 years old.

Figure 13. Line of animal heads engraved on a rib from the cave of Courbet, Penne-Tarn, France, Late Magdalenian, about 12,500 years ago (source: Wikimedia Commons)

The patchy nature of the expression of autistic traits in post 100,000 bp archaeological evidence suggests that it may have been most particularly in certain times and places that the advantages of autism were particularly emphasised. The clearest context may be that of highly risky, cold climate environments such as Ice Age Europe. Here both the dependence on technology for survival is greatest, and technological efficiency and innovation much valued, and unstable climates place an emphasis on large scale collaborations to provide a social buffer against shortfalls in resources. The archaeological record of southernmost Africa provides another particularly interesting case where autistic traits appear to be have adopted early at Blombos and Rose Cottage, later declined around 65,000 years ago and then re-adopted many thousands of years later continuing into elements of modern San technology (see [51]). If those with autism were integrated into societies at different times and places it is not surprising the there are multiple genes coding for autistic traits, and representing a geographically complex process of selection.

3.1. The cloud behind the silver lining?

Individuals with autism can create challenges for societies, whether small scale hunter-gatherers or large scale modern societies. Pronounced counter-dominance tactics in hunter-gathers (see [52]) for example may have developed in part to prevent the dominance of those with autistic traits such as rigid rules and a lack of sensitivity to potential emotional consequences of their actions. Thus no matter how much someone is respected in small scale egalitarian groups, their rights to dictate the behaviour of others is heavily constrained by shared action to maintain equality. Indeed Boehm documents a progressive series of sanctions for dominating behaviour from ridicule to ostracism or assassination (see [53]). Such dynamics have also been recognised in Palaeolithic and Mesolithic contexts (see [52]). Whilst counter-dominance tactics work in a small scale setting, in modern societies a lack of such intuitively based social sanctions on behaviour may create problems where highly dominant individuals with autism are in positions of power. In this case such individuals may make decisions with emotionally damaging consequences for others which remain unchallenged.

Most individuals with autism are highly moral. However where autism is associated with disorders of *motivation*, as in the case of autistic psychopathology, a lack of intuitive feeling of others' suffering allied with a desire to harm can be a literally lethal combination (see [54]).

Whilst we tend to envisage a single typical Upper Palaeolithic society, such societies, like the human minds within them, were likely to have been highly variable. In a modern context small scale societies today tend to vary greatly in their level of social tolerance and tendencies towards or against violence, and a certain amount of self sorting takes place amongst hunter-gatherers with more collaborative or more competitive individuals tending to group together (see [55]). It is not difficult to envisage situations in prehistory where it was not the highly collaborative and moral personalities which were the most successful but in contrast where highly dominant, aggressive and even violent attitudes towards others might occasionally 'pay off' sufficiently to allow the genetic determinants of such traits to be selected for (see [54]).

3.2. The timing of incorporation of autism

Why might the incorporation of autism apparently occur relatively late in human evolution (at least after 100,000 years ago)? A capacity to integrate those who think differently, not only at the autistic end of the spectrum but also by implication other differences in mind, may depend on particular evolutionary changes taking place. Perhaps the most likely is that a particular cognitive threshold might need to be passed. A capacity to care about and support members of society is in evidence at much earlier dates. However we might speculate that only when early humans had the cognitive sophistication to appreciate that *behind different behaviour lies positive motivations towards others* as well as *the moral consciousness to promote inclusivity* could

http://dx.doi.org/10.5772/53883

autism bring the unique elements to make up 'humanity'. These unique elements may nonetheless have been a critical part of the remarkable global colonisation and modern human success which follows their appearance.

4. Conclusions

There is every reason to believe that autism, far from being outside society, is very much part of the story of the origins of 'humanity'. Those with autism may have played a unique role in technological spheres and understanding of natural systems, contributing calendrical knowledge, refined efficient technological practices and a unique perspective to art. They may also have been key to allowing larger scale societies to form with clear rules to define how sharing takes places.

Autism is sometimes portrayed as the 'other'. Not only is this a dangerous perspective to take on a difference of mind, but there is every reason to conclude that autism is a central part of what makes us 'human'. However difficult dealing with autism may be there may be much which we owe to the role of autism in our success. Moreover the solutions to allowing 'us' to work with 'others' in the Palaeolithic, and allowing a large scale society to be created might have depended on the inclusion of autism.

Acknowledgements

I much appreciate lively discussions and advice not only from my undergraduate and postgraduate students but also from colleaugues, most particularly Barry Wright, Andy Needham, Isabelle Winder, Geoff Bailey, Mark Edmonds, Andy Shuttleworth, Adam Feinstein, Paul Trehin and Nicolas Humphrey. All errors are my own.

Author details

Penny Spikins

Department of Archaeology, King's Manor, University of York, UK

References

[1] Grinker R R. Commentary: On being autistic and social, *Ethos* 2010; 38 (1) 172-8.

[2] Barkow J, Cosmides L, Tooby J. *The Adapted Mind: Evolutionary psychology and the generation of culture.* NY: Oxford University Press; 1992.

[3] Walker A, Zimmerman M R, Leakey R E F. A possible case of hypervitaminosis A in *Homo erectus*. *Nature* 1982; 296: 248-250.

[4] Spikins P A, Rutherford H, Needham A. From hominity to humanity: compassion from the earliest archaics to modern humans. *Time and Mind* 2010; 3 (1) 303-325.

[5] Saladié P H, Rodríguez-Hidalgo R, Cáceres A, Esteban-Nadal I, Arsuaga M, Bermúdez de Castro J- L, Carbonell J.-M. Intergroup cannibalism in the European Early Pleistocene: The range expansion and imbalance of power hypotheses, *Journal of Human Evolution* 2012, Available online 31 August 2012, ISSN 0047-2484, 10.1016/j.jhevol.2012.07.004 (http://www.sciencedirect.com/science/article/pii/S0047248412001406)

[6] Xu X, Zuo X, Wang L, Han S. Do you feel my pain? Racial group membership modulates empathic neural responses. *Journal of Neuroscience* 2009; 29: 8525–8529.

[7] Cikara M, Botvinick, M M, Fiske, S T. Us versus them: social identity shapes responses to intergroup competition and harm. *Psychological Science* 2011; 22: 306– 313.

[8] Marwick B. Pleistocene exchange networks as evidence for the evolution of language. *Cambridge Archaeological Journal* 2003;13: 67-81.

[9] Whitley D S. *Cave paintings and the human spirit: The origins of creativity and belief.* Prometheus Books; 2009.

[10] Spikins P A. Autism, the integration of difference and the origins of modern human behaviour. *Cambridge Archaeological Journal* 2009; 19 (2)179-201.

[11] Baron-Cohen S. *Zero degrees of empathy: A new theory of human cruelty and kindness.* Penguin; 2012.

[12] Jaarsma P, Welin S. Autism as natural human variation: reflections on the claims of the neurodiversity movement, *Health Care Analysis* 2012; 20 (1) 20-30.

[13] Ochs E, Solomon O. Autistic sociality, *Ethos* 2010; 28 (1) 69-92.

[14] Baron-Cohen S. Is asperger syndrome/high functioning autism necessarily a disability? *Development and Psychopathology* 2000; 12: 489-500.

[15] Baron-Cohen S. Does autism need a cure? *The Lancet* 2009; 373: 1595-1596.

[16] Myers P, Wheelwright S. *An exact mind: An artist with asperger's syndrome.* Jessica Kingsley; 2004.

[17] Baron-Cohen S, Ashwin E, Ashwin C, Tavassolit T, Chakrabarti B. Talent in autism: hyper-systemising, hyper-attention to detail and sensory hypersensitivity. *Philosophical Transactions of the Royal Society of London B. Biological Sciences* 2009; 364: 1377-83.

[18] Baron-Cohen S, Bolton P, Wheelwright S, Scahill V, Short L, Mead G, Smith A. Autism occurs more often in families of physicists, engineers and mathematicians. *Autism* 1998; 2: 296-301.

[19] Rodman K E. *Asperger's Syndrome and Adults ... Is Anyone Listening?* (FAAAS Inc.) London: Jessica Kingsley; 2003.

[20] Dziobek I, Rogers K, Fleck S, Bahnemann M, Heekeren H, Wolf O, Convit A. Dissociation of cognitive and emotional empathy in adults with asperger syndrome using the multi-faceted empathy test (MET). *Journal of autism and developmental disorders* 2008; 38: 464-473.

[21] Solomon O. Sense and the senses: anthropology and the study of autism. *Ethos* 2010; 38 (1): p245

[22] Sacks O. *An Anthropologist on Mars: Seven Paradoxical Tales*. New York: Knopf; 1995.

[23] Gernsbacher M A, Dawson M, Mottron L. Autism: common, heritable, but not harmful. *Brain and Behavioural Sciences* 2006; 29(4): p14.

[24] Pickover C A. *Strange Brains and Genius: the Secret Lives of Eccentric Scientists and Madmen.* London: Harper Perennial;1999.p35.

[25] Fitzgerald M. *Autism and Creativity: Is there a link between autism in men and exceptional ability?* Brunner Routledge: New York; 2004.

[26] Fitzgerald M. *The Genesis of Artistic Creativity: Asperger's syndrome and the Arts.* London: Jessica Kingsley; 2005.

[27] Walker A, Fitzgerald M. *Unstoppable Brilliance*. Dublin: Liberties Press; 2006.

[28] Fitzgerald M, O'Brien B.*Genius Genes: How Asperger's syndrome Changed the World.* Autism Asperger Publishing Company; 2007.

[29] Prince D E. An exceptional path: an ethnographic narrative reflecting on autistic parenthood from evolutionary, cultural and spiritual perspectives. *Ethos* 2010; 38 (1): p62.

[30] Baron-Cohen, A. The hyper-systemising assortative mating theory of autism, *Progress in Neuro-Psychopharmacology and Biological Psychiatry* 2006; 30 (5) 865–872.

[31] Turner-Brown L, Lam K, Holtzcla T N, Dichter G S, Bodfish J. Phenomenology and measurement of circumscribed interests in autism spectrum disorders, *Autism* 2011; 15 (4) 437-456.

[32] Spiker M A, Enjey Lin C, VanDyke M, Wood J. Restricted interests and anxiety in children with autism. *Autism* 2012; 16: 306.

[33] Williams E, Costell A. Taking things more seriously: Psychological theories of autism and the material-social divide, in P. M. Graves-Brown (ed.) *Matter, Materiality and Modern Culture*. Routledge. London; 2000.

[34] Selfe L. *Nadia Revisited: A longitudinal study of an autistic savant.* Psychology Press; 2011.

[35] Kellman J. Ice age art, autism and vision: How we see/How we draw, *Studies in Art Education. A journal of issues and research* 1998; 39 (2) 117-131.

[36] Batson C D, Klein T R, Highberger L, Shaw L L. Immorality from empathy induced altruism: When compassion and justice conflict. *Journal of Personality and Social Psychology* 1995; 68 (6) 1042-1054.

[37] Balicko A. *The Netsilik Eskimo.* Waveland Press; 1970: p133.

[38] Brantmöller M, Pastoors A, Weninger B, Weninger G-C. The repeated replacement model – rapid climate change and population dynamics in Late Pleistocene Europe. *Quaternary International* 2012; 247: 38-49.

[39] Green R E, Krause J, Briggs A W, Maricic T, Stenzel U, Kircher, Petterson N, Li H, et al. A draft sequence of the Neanderthal genome. *Science* 2010; 328: 710-722.

[40] Meyer M, Kircher M, Gansuage M-T, et al. A high coverage genome sequence from archaic Denisovan individual. *Science* 2012; 1224344. Published online 30 August 2012 [DOI:10.1126/science.1224344]

[41] D'Errico F, Stringer C. Evolution, revolution or saltation scenarios for the emergence of modern cultures. *Philosophical Transactions of the Royal Society B* 2011; 366: 1060-1069.

[42] Shea J J, Sisk M L. Complex projectile technology and Homo sapiens dispersal into western Eurasia. *PaleoAnthropology* 2010;100–122.

[43] Marshack A. The Taï plaque and calendrical notation in the upper Palaeolithic, *Cambridge Archaeological Journal* 1991; 1 (1) 2.

[44] Henshilwood C, Dubreuil B. The Still Bay and Howieson's Poort: Symbolic material culture and the evolution of mind in the African Middle Stone age, *Current Anthropology* 2011; 52 (3) 361-400.

[45] Mellars P. Going East: New genetic and archaeological perspectives on the modern human colonization of Eurasia. *Science* 2006; 313: 796-800.

[46] Hayden B, Villeneuve S. Astronomy in the Palaeolithic? *Cambridge Archaeological Journal* 2011; 21: 331-355.

[47] Jegués-Wolkiewiez C. Aux racines de l'astronomie ou l'ordre caché d'une oeuvre paléolithique. *Antiquités Nationales* 2005; 37: 43–52.

[48] De Smedt J, De Cruz H. The role of materialculture in human time representation. Calendrical systems as extensions of mental timetravel. *Adaptive Behaviour* 2011; 19, 63–76.

[49] Humphrey N. Cave art, autism and the evolution of the human mind, *Cambridge Archaeological Journal* 1998; 8: 165-191.

[50] Trehin P. *Palaeolithic art and autistic savant syndrome,* Presentation in 7th International Congress, Autism Europe, Lisbon November 14-16th ; 2003.

[51] D'Errico F, Blackwill L, Villa P, Degano I, Lucejko J, Bamford M, Higham T, Colombini M P, Beaumont P. Organic artefacts from Border Cave: earliest evidence of San material culture, *ESHE* 2012; 1: 65.

[52] Spikins P A. The bashful and the boastful: prestigious leaders and social change in Mesolithic societies. *Journal of World Prehistory* 2008; (3-4): 173–93.

[53] Boehm C. *Hierarchy in the forest, the evolution of egalitarian behaviour*. Cambridge MA: Harvard University Press; 1999.

[54] Fitzgerald M. *Young, violent and dangerous to know.* Nova Publishers; 2010.

[55] Apicella C, Marlowe F W, Fowler J H, and Christakis N A. Social networks and cooperation in hunter-gatherers, Nature 2012; 481: 497-501.

Section 2

Attitudes to Autism

Chapter 2

Attitudes Towards Autism Among Israeli Arab Teachers' College Students

Karni-Vizer Nirit and Reiter Shunit

Additional information is available at the end of the chapter

http://dx.doi.org/10.5772/54845

1. Introduction

The rise in the prevalence of autistic spectrum disorder has increased world-wide (Flood, Bulgrin and Morgan, 2012). This phenomena calls for the expansion and development of teachers preparation courses to work with this population as well as providing specific courses to professionals in the medical and health professions such as speech therapists, occupational therapists, art therapists etc. that focus on the needs of children and adults with autism. It also calls for the development of community services to children and families, such as early diagnosis, counseling parents, education of school aged children and as they graduate from school, creating community living arrangements for adults along with opportunities for employment, social and leisure time activities. The development and use of appropriate services must be based on awareness of the autistic spectrum disorder as a subgroup of persons with developmental disabilities that is distinguished from other disabilities such as intellectual disabilities or mental illness. It is also based on openness and positive attitudes towards these persons as equal members of society who have the right to live in the community in a respectful manner.

The present chapter will describe a survey done of the attitudes regarding persons with autism in the Israeli Arab community, according to Arab students attending an Arab teachers' training college.

Future teachers were chosen for the study as leaders of the young generation who will be the citizens of the future. The aim of the study was to assess the current views and attitudes among would be teachers in the Israeli Arab community regarding the autistic spectrum disorder so that we can develop suitable teachers' preparation courses along with suitable services for the families population and for children and adults with autism.

The importance of preparing teachers in the Arab world in the area of autism is emphasized by Al-Shammari Zaid (2006) in his paper on *Special Education Teachers' Attitudes Toward Autistic Students in the Autism School in the State of Kuwait.* He states that though a special school for children with autism was established in 1999 in the State of Kuwait, the Kuwaiti universities which prepare future teachers lack courses and curricula to help pre-service and in-service special teachers understand issues related to autism. According to Al-Shammari Zaid teachers, are ill-prepared to teach students with autism in Kuwait. Also, since the diagnosis of autism is relatively new to Kuwait, teachers are not fully equipped on how to best teach students with autism.

Lack of teachers' preparation coursed and professional know how is becoming acute in the Arab world since in the past decade there has been a constant increase in family health care. In Saudi Arabia for example, Amr Mostafa (2011) in his paper on *Addressing autism in the Arab world* notes that "In the Middle East and North Africa there has been a fivefold decrease in child mortality in recent decades."

Also, the high rates of prevalence of Autism in the Arab world may be due to genetics. For example, in Saudi Arabia almost one third of children with autism were born to parents who are close relatives, usually first cousins. This is dependent on location – higher in rural communities (up to 80%) and lower in urban and suburban settings (34%). This phenomenon suggests that families in Saudi Arabia have a higher incidence of autism. Inter related family marriages are actually the norm in most Muslim countries so that this might result in a higher prevalence in other Arab nations as well.

In spite of this, Amr Mostafa (ibid) notes that the condition of autism has long been hidden in much of the Arab world. Furthermore, he also suggests that it is often difficult for a child with autism to be diagnosed properly, because of poor professional knowledge in this area. Pediatricians are relatively inexperienced in the diagnosis and management of psychiatric disorders including children with autism and there are few psychiatrists specializing in childhood development problems. He also notes the lack of awareness among parents, a failure to recognize symptoms and seek diagnosis and treatment. His conclusion is that both under-diagnosis and under-reporting may play a role in the disparity in prevalence and consequently, scarcity of appropriate services.

A similar picture is portrayed by Hanan, Elbahnasawy and Naglaa, Girgis (2011) from Egypt. In their paper "Counseling for Mothers to Cope with their Autistic Children", they point to the fact that there has been a dramatic increase in the incidence of autism from the 1970's through 2008 – the prevalence of autism in Egypt is increasing, with one child out of every 870 Egyptian children being diagnosed as having autism.

Apart from lack of professional knowhow, the authors add another factor that hinders the appropriate care of families and children in the Arab world. According to them, caring for a family member with autism costs money and places a burden on family finance. A recent study in Egypt found that 83.3% - 91.3% of people with autism live at home with their families well into their adult years since there are no appropriate services for them. Autism care in Egypt is usually home-based with less than a quarter being enrolled in schools. They are informally

http://dx.doi.org/10.5772/54845

taught by parents, siblings, grandparents or friends. Some non-profit organizations such as Arab autism societies, try to address this problem by offering specialized education.

In addition to financial difficulties that can be a barrier to appropriate care, cultural beliefs and attitudes can be a hindrance too. There are two interrelated attitude sources towards persons with disabilities, the religious source and common prejudices held by the general public and transmitted from one generation to the next in the form of stereotypes (Reiter, 1999).

Estela Dimes (2012) a Filipino American married to a Jordanian American have a child with autism and she describes the way cultural traditions, values and beliefs can affect how people deal with this disability. She compares two different views. According to her, in the Filipino culture, having a child with a disability is viewed in a positive way even as "a blessing or gift from God". In comparison Middle Easterners, especially among the older-generation, believe that a disability is a form of God's punishment for sins or perhaps the result of a curse. She continues to describe the attitudes of 'Middle Easterners' as exhibiting shame and embarrassment. Because of that the author assumes that they tend to overlook developmental differences in their children with autism trying to avoid the stigma attached to it. She concludes that this social attitude often causes delay in diagnosis, until the child is of school age, thus preventing early intervention crucial to treatment of developmental disorders.

This fatalistic attitude is also addressed by Al-Shammari Zaid (2006) who further describes the cultural beliefs about disability in the Arab society. Actually, he observed that caring for a child with autism can bring the family closer to God. This is based on the belief in Karatma, God's Will – or one's destiny. This belief is often invoked to provide an explanation to major life events, including the occurrence of disability. Like Estela Dimes (2012) he also suggests that an explanation for the occurrence of a disability in the family is the belief that it is a result of sinful behavior committed in the past. Thus, the family tends to accept their own disabilities, as fate due to God's will.

One direct outcome of this attitude is that since it was God's will that the child was born with a disability his life is in the hand of God. For this reason, Al-Shammari Zaid (2006) considers the most important problem when working with families in the state of Kuwait, is that even when a child is enrolled in a special education individualized program his family does not apply it at home but rather, continue to isolate the child in his personal life.

The other source of attitudes towards persons with disabilities are prejudices held by the public in the form of stereotypes. According to The American Heritage® Dictionary of the English Language, (Fourth Edition copyright ©2000) a prejudice is:

1. **a.** An adverse judgment or opinion formed beforehand or without knowledge or examination of the facts.

1. **b.** A preconceived preference or idea.

2. The act or state of holding unreasonable preconceived judgments or convictions.

3. Irrational suspicion or hatred of a particular group, race, or religion.

4. Detriment or injury caused to a person by the preconceived, unfavorable conviction of another or others.

Prejudices are expressed in cultural stereotypes affecting the discrimination of those seen as belonging to the group in question. Stereotypes obviously affect social judgments we make about others. They influence how much we like the person.

Prejudice is an attitude toward the members of some group based solely on their membership in that group (can be positive or negative), while stereotypes involve generalizations about the "typical" characteristics of members of that group. Prejudice and stereotypes lead to discrimination, i.e. the actual positive or negative actions toward the objects of prejudice. Children acquire negative attitudes toward various social groups through direct and vicarious learning experiences. Parents, teachers, friends, the mass media all play roles in the development of prejudice. The fact that prejudice is a learned reaction, unables the process of change – the unlearning of old views and attitudes and the relearning of new ones. Since stereotypes distort our perceptions and affect the information we note - we'll give more attention to stereotype-consistent information and stereotype-inconsistent information that activates attempts to disconfirm/reject, by imparting new knowledge concerning the group in question, we can introduce change in attitudes towards that group and achieve a breakdown of the stereotypes held about them. Thus we can reduce prejudice through cognitive interventions such as teaching and providing new information that is undistorted and scientifically based. For example, research on attitudes change towards persons with disabilities among Arab students who attended special education courses as part of their professional development, showed significantly more positive attitudes towards persons with disabilities than their colleagues who did not have such a background (Devine, 1989, Karni, Reiter, Bryen, 2011).

2. Arabs in Israel

The Arab population in Israel today numbers about 1.4 million residents, who comprise about 20% of the country's total population of approximately 8 million people. Among the Arab population of

Israel, Moslems comprise about 82.5%, Christians about 9%, and Druze about 8.5% (as of 2004). Most of Israeli Arab population live in the northern part of the country – the Galilee. They reside in small and large, purely Muslim, Christians or Druse villages, as well as in mixed towns.

Most of the Bedouins live in the Southern region of the country.

3. Prevalence of people with disabilities in the Arab sector in Israel

Notwithstanding the difficulty involved in obtaining accurate quantitative data on people with disabilities in the Arab sector there are indications that the percentage of people with disabilities is higher in the Arab than in the Jewish sector. This can be attributed to several causes, including the high rate of interfamily marriages, hereditary diseases, and women giving birth at a relatively late age (among Bedouins, for example) and a general lack of awareness of various genetic problems.

In the last decade, Arab society in Israel has undergone continuous change, and as a result of the modernization process, a transition occurs from a collectivistic society to a more nuclear one. There is an increase in the level of education and an increase in the number of women joining the work force. These two major social factors cause changes in the social structure and social norms and values of Israeli Arab population. In spite of these changes, the traditional bonds still exist and have a strong hold on the individual. Traditionally the 'Hamula (Clan)' was the most binding collective social organization. As an extended family system, it provided economic social and psychological security for its members. It is a patriarchal system: fathers always have the upper hand especially in decision making and socialization of children in the family; women tend to be passive and conforming. Ideally, children are to be obedient and conforming too (Odeh, 2007, Al Haj, 1989, Mar'i, 1978). As Mar'i and Karayanni (1982) observed: the modernization process traditional structures and the values associated with them have weakened: and in their place new structures are evolving alongside the adoption of more modern patterns of social organization and new values and behaviors. However, this change has not been a smooth one or without inherent contradictions. Quite the opposite, contradictions and even clashes between the old and the new are so common as to suggest that the society is conflict-ridden in all that concerns conservation and change, continuity and discontinuity.

According to a comprehensive research report done in Israel concerning people with disabilities in Arab society in Israel (Sandler-Loeff, Shahak, 2006) they have to cope with numerous barriers. Some of the barriers are similar to those faced by all disabled persons, whereas others are specific to or exacerbated by the Arab socio cultural context. Surveys done in Israel (ibid) found that the attitudes towards people with disabilities in Arab society, show lack of knowledge and lack of services; lack of access for disabled individuals; a societal attitude that does not accept them as having equal rights; and lack of coordination between the various agencies that deal with the affairs of people with disabilities.

However, the past decade in Israel at large has witnessed important developments in the lives of people with disabilities, and they are also evident in Arab society. These developments can be largely attributed to changes in legislation and to the development of services.

However, Arab parents show a greater tendency than Jewish parents to be, they are overprotective of their disabled children and give these children less encouragement to be independent, make decisions and find employment. Arab parents of children with disabilities reported a greater feeling of shame and hypersensitivity to the reactions of others than did Jewish parents. Indeed Israeli Arab attitudes towards autism spectrum disorder are not unlike the ones found in the neighboring Arab countries. Religion has a powerful spiritual effect. It is both a source of comfort for Arab families and at the same time a barrier to 'normalization' and efforts to include the child in regular society. This was evident in a study done in Israel of Druze mothers to children with Autism (Al-Shich, 2012). Al-Shich, herself a Druze conducted in depth interviews with 10 mothers who live in the villages in the northern part of Israel. The major theme that emerged was the strong religious beliefs attached to the disability of the child.

In Israel there are approximately 122,000 Druze. As indicated before concerning intermarriages in the families, here too intermarriage is high and reaches 47% of all marriages (Vardi-Saliternik, Friedlander, Cohen, 2002). They represent 8.3% of the Arab population in the country. They live in 18 villages in the Northern parts of the Israel, the Carmel Mountain, the Galilee and the Golan Heights (2009). The Druze community is a traditional society with a collectivistic orientation emphasizing the supremacy of the extended family and the community over the individual.

Though they are Muslims, they have a unique religious orientation. The core of their religious principles are known only to a small number of religious persons and are kept secret not only from outsiders but also from the common Druze. All Druze believe in incarnation according to which life on earth is temporary, once dead the spirit will be reincarnated in another body. This belief promotes an optimistic view of life since if you are not successful or handicapped in your present life, you can be very successful and with no disabilities in your next life (Kandel, Morad, Vardi, Press, & Merrick,2004). Indeed, in a research done with Israeli Arab parents of children with developmental disabilities by Reiter, Mar'i and Rosenberg, (1986) comparisons between Muslim Arabs, Christian Arabs and Druze parents revealed that attitudes towards the child were most positive among the Druze. Indeed, in Meissa's study (ibid), eight mothers out of ten, said that having a child with Autism is God's Will and should be regarded as a trial of faith. One mother described it as a gift from God and one said it was a punishment for sinful behavior. Four mothers said that having a child with autism made them become more religious, or 'closer to God', one family actually converted from being atheists to believers.

With this recognition of the strong religious beliefs attached to disability, a special program was designed in Israel to give training to Imams – the Muslim religious leaders, to change perceptions of mental illness, including autism in Arab Israeli society. The assumption underlying the program was that religious figures are in positions of influence in their communities. The public sees them as societal and spiritual guides and turn to them for advice and guidance. They are also figures of authority in their communities and can guide families in times of distress. As such they can act as agents of positive change (Imans receive training to change perceptions of mental illness in Arab Israeli society, Ynet News, Monday, November 22, 2010).

Some 40 imams from mosques in Israeli Arab communities participated in a unique training program for changing perceptions of mental illness in Arab Israeli society. Five weekly meetings dealt with issues like the religious aspects, ways of coping, getting to know persons with disabilities and their families, legal issues and service, the effects of medication, mental illness from a medical point of view and ideas for social change.

Prof. Hawala Abu Bakar from Emeq Yezreel College and Al-Qasemi Arab Teachers College who specializes in mental illness in Arab Israeli society comments that there is a gap between

http://dx.doi.org/10.5772/54845

the acceptance and compassion that the religions (Islam, Christianity and Druze) demand towards those with disabilities and the reality on the ground.

Attitudes range between two extremes – acceptance, which can even include encouragement to marry, and ostracism, especially towards those with rare deformities or those who are violent towards themselves or the environment. There is also a difference between attitudes towards men and women, with the tendency to isolate women more than men.

In Israel The most widely treated disability in Arab localities is mental retardation, which is given considerable attention, followed by sensory impairments (blindness and deafness), and physical impairments (handicaps and illnesses). Impairments and disabilities in the area of mental health, various behavioral, communication, and functional disorders and autism, are given relatively little attention and very few agencies deal with those areas.

The authors of the special report on people with disabilities in Arab Society in Israel (ibid) conclude that the lack of diagnostic capabilities together with insufficient knowledge and awareness about developmental problems in Arab society create a situation in which disabilities are insufficiently diagnosed, or that diagnosis occurs too late when early diagnosis could have prevented further disability or deterioration. There is inaccessibility and unavailability of mental health clinics in the Arab sector, coupled with language and cultural barriers within mental health clinics since most service providers are Jewish. Another explanation is that in Arab society there is insufficient acknowledgment or awareness of the benefits of psychological care and mental health services, as well as a preference for relying upon informal local resources within the community.

Assistance from formal agencies or organized efforts on the local level are a new developments seen over the past decade and are not yet firmly rooted in the society. The process of transition from a culture reliant on the nuclear and extended family as a pillar of society to a culture which promotes civil society, including initiatives on behalf of the disabled, has been gradual and difficult.

In recent years though there has been increasing awareness of the need to integrate people with disabilities into the community in the Arab sector (Karni, Reiter, and Bryen, 2011). It was the aim of the present survey to focus our understating on the way future Arab teachers attending an Arab teachers' college perceive 'Autism' and their attitudes towards persons, children and adults, with autism. Our secondary aims were to find out to what extent: previous contact, or having a family member with autism, affected attitudes;

- gender has an impact;
- age is a crucial factor;
- attending, or having attended a course(s) on autism and other developmental disabilities affected attitudes.

The first hypothesis related to attitudes toward the inclusion of persons with autism and the belief that they can achieve social and academic skills that will enable them to be socially included in the community.

The first hypothesis stated that there will be correlations between positive attitudes in the areas of knowledge about autism and emotional and behavioral willingness to be close to them, and positive conceptions of their abilities in the areas of life skills, academic skills and included social competence.

The second hypothesis related to **knowledge** about the syndrome.

The hypothesis stated that those who attended a course on autism will show better knowledge of the syndrome than those who did not.

The third hypothesis stated that there will be positive correlations between the background variables of previous contact and/or having a family member with autism and better knowledge of and positive attitudes towards them.

The forth hypothesis was divided into two parts: a. religion will be pointed out as the major factor underlying attitudes and conceptions regarding autism, more so than prejudice or personal experience with a person with autism, and b. religion will be correlated with the most negative attitudes and conceptions, more than prejudice and more than personal experience.

4. Method

4.1. Sample

Students attending the Sakhnin's teachers education college were asked to fill in two questionnaires on their attitudes towards Autism. The sample included 321 students from all study levels, from the first to the fourth year of study. There were 82.3% females and 17.7% males. The mean age was 26.58 (range: 19 – 49). The wide age range was due to the fact that some students attend a special program for veteran teachers who wish to get a degree in special education.

4.2. Instruments

Two questionnaires were used in this study. One was an especially designed questionnaire including two parts: background variables such as: personal knowledge of a person with autism, autism in the family, heard about autism before studying from the mass communication media or from other people, cultural attitudes toward autism (positive, negative, indifference), reasons for cultural attitudes (religion, prejudice, personal experience), autism being different from intellectual disability (yes/no), the existence of services for children and adults with autism (yes/no). In the second part of the questionnaire the respondent was asked about his/her perceptions of the possibility of inclusion of persons with autism based on their ability to be independent by developing appropriate life skills, have the academic abilities for inclusion and being able to be socially included in the community. The total reliability score of the scale was high (Cronbach's α =.84); life skills sub scale (Cronbach's α =.78); academic skills sub scale (Cronbach's α =.69); social inclusion sub scale (Cronbach's α =.71).

The second questionnaire was the Hebrew translation of the Chedoke-McMaster Attitudes Towards Children with Handicaps (CATCH) scale (Rosenbaum et al., 1986; Tirosh, Schanin, Reiter, 1997). The questionnaire is a 28 – item self-administered scale scored on a Likert 4 point scale from 1 = always true to 4 = never. It was adapted to refer to the autistic spectrum disorder rather than disability in general. Items were worded as statements, for example: people with autism cannot be happy in their lives; I have no problem spending my leisure time with a person with a autism; I am ready to talk with a person with autism whom I do not know; I feel uneasy introducing a person with autism to my friends.

Factor analysis (Factor analysis with varimax rotation and the criterion of Eigenvalue greater than 1) revealed three sub-scales: willingness to be close to a person with autism (Cronbach's α =.83); emotional reactions to a person with autism (Cronbach's α =.69); knowledge of autism (Cronbach's α =.51). The last sub-scale of knowledge has a rather low reliability score showing the fragmentation of knowledge respondents have in this area. The three subscales are in line with the three dimensions of attitudes, the emotional, the behavioral and the cognitive suggested by Reiter and Bryen (2010). Accordingly, attitudes are composed of three interrelated dimensions of personality: emotions, cognition and behavior. Each dimension is a complex interplay of several factors. Looking at the cognitive dimension, an attitude is a view or opinion that a person has towards a certain state of existence, of an object, an idea, of another person, or of other people. Attitudes are frequently swayed by emotional response. They can be positive, for example, happiness, pleasure, wanting to experience an event, be near the person, or get hold of the object of reference. Attitudes can also be negative, when a person feels he has a dislike towards a situation, object, or another person. In this case the person will likely feel unhappy, fearful, disgusted, sad, etc. The two dimensions of cognition and emotion may not always be congruent. The third dimension of attitudes is a behavioral one. Incongruence between cognition, emotion and behavior can be the outcome of cultural norms and commonly held prejudices.

The two questionnaires were administered to students during class time.

4.3. Results

The analysis of the background variables of our sample revealed that 66.5% of the total sample, took courses in special education including a course on autism, as compared to 33.5% who had not taken part in any academic studies in the area of disabilities. Most of those who took the course (84.1%) said that it changed their attitudes towards autism in a positive way. Before attending the course, most respondents did not know that autism was a different syndrome than intellectual disability (61.6%).

Nearly one half (43.3%) knew a person with autism, while only a very small number (7.2%) said there was a person with autism in their family. Asked about their opinion as to the reason for the common attitudes towards autism in the Arab community respondents pointed to prejudice as the main source of views (66.3%) others thought it was based on personal experience of people with persons with autism (22/1%) and some thought they were due to a religious outlook (11.6%).

Asked about the existence of services for children and adults with autism, as far as they knew, more than one half (53.7%) said there were no services for adults and only slightly less than half (43.1%) said there were no Arab services for children. It should be noted though, that in relations to services for adults, indeed there is a scarcity of community living arrangements and sheltered workshops for them. However, regarding school aged children, all children in Israel, irrespective of any disability, must attend school by law. Thus, all school age children, including those with autism, from the age of 3 to 21 (for children with disabilities) attend either an Arab school or a Jewish school. It can be a special education school or a regular school, depending on the special needs of the child and parental wishes regarding the placement of the child. When indicating that there is lack of schools for the younger age, respondents referred to special education Arab schools for children with autism.

The first hypothesis stated that there will be correlations between positive attitudes in the areas of knowledge about autism and emotional and behavioral willingness to be close to them, and positive conceptions of their abilities in the areas of life skills, academic skills and included social competence.

Applying the Spearman Correlations test on scores obtained on the two questionnaires administered in the study, significant correlations were found between all the sub-scales of the questionnaires i.e. positive attitudes in the three areas of knowledge, behavior and emotions; correlated positively and significantly with perceptions of their ability to acquire life skills, academic skills, and social skills.

Table 1 presents the correlations obtained between attitudes and perceptions of skills.

		Attitudes				**Perceptions / knowledge**			
		Total attitudes	Closeness	Emotions	Cognitive	Total perceptions	Life skills	Academic skills	Social integration
Attitudes	Total attitudes		.90***	.77***	.48***	.39***	.24***	.29***	.39***
	Closeness			.60***	.16**	.37***	.20***	.27***	.41***
	Emotions				.26***	.15**	.03	.18**	.18**
	Cognitive					.28***	.28***	.22***	.18**
Perceptions / knowledge	Total perceptions						.82***	.68***	.87***
	Life skills							.38***	.52***
	Academic skills								.43***
	Social integration								

*p<.05, **p<.01, ***p<.001

Table 1. Spearman Rank Order correlations between full attitude scale and its sub scales and full perception scale and its subscales for the total sample (N= 321)

http://dx.doi.org/10.5772/54845

As seen from table 1 correlations were highly significant (p<.001) between the full scales scores and all sub scales of the questionnaires, indicating that positive attitudes towards persons with autism correlated with positive perceptions regarding the possibility of their inclusion in the community.

The second hypothesis related to **knowledge** about the syndrome.

The hypothesis stated that college students who attended a course on autism will show more positive attitudes towards persons with autism and more positive conceptions of their potential for inclusion than those who did not attend such a course. Furthermore, on a general question asking the students who attended a course on autism whether it changed their attitudes, most of them (84.1%) said that it did change their attitudes in a positive way.

Table 2 represents t-tests between scores obtained on the questionnaires by the students who attended a course on autism compared with those who did not attend such a course.

		Yes (n=211)	No (n=106)	F(1,315) (η^2)
Attitudes	Total attitudes	3.03 (0.38)	2.90 (0.43)	7.41** (.023)
	Closeness	3.19 (0.49)	3.02 (0.56)	8.04** (.025)
	Emotions	3.37 (0.54)	3.26 (0.62)	2.77 (.009)
	Cognitive	2.45 (0.51)	2.43 (0.53)	0.10 (.001)
Perceptions / knowledge	Total perceptions	2.59 (0.48)	2.54 (0.52)	0.55 (.002)
	Life skills	2.28 (0.65)	2.28 (0.69)	0.01 (.001)
	Academic skills	2.90 (0.68)	2.81 (0.76)	1.03 (.003)
	Social integration	2.67 (0.52)	2.61 (0.56)	0.90 (.003)

*p<.05, **p<.01, ***p<.001

Table 2. Multivariate analyses of variance between the students who attended a course on autism (N= 211) and those who did not attend such a course (106) on the full attitude scale and its sub scales and full perception scale and its subscales

From table 2 it appears that attending a course on autism significantly (p<.01) affected the total attitudes scale, especially the willingness of the respondents to have close relationships with a person with autism. However, the course did not make a difference on students' perceptions of the possibility of persons with autism to be included in the community.

Apparently, as the saying goes 'old attitudes die hard'.

The third hypothesis stated that there will be positive correlations between the background variables of previous contact and/or family contact with a person with autism and better knowledge and positive attitudes towards them.

Multivariate analyses of variance on the variable of *'I know a person with autism'* and the variable of *'there is a person with autism in my family'* revealed no significant differences between those who knew or had a relative with autism and those who did not know or did not have such a relative.

Thus, our third hypothesis was not confirmed.

The forth hypothesis stated that among the factors underlying attitudes and conceptions regarding autism, religion will be pointed out as the major factor, more so than prejudice or personal experience with a person with autism, religion will be correlated with the most negative attitudes and conceptions, more than prejudice and more than personal experience.

Comparisons using Multivariate analyses of variance with Tukey's post hoc tests were applied in order to find out which among these variables was indicated by most respondents as the major factor underlying attitudes and conceptions and which affected mostly negative attitudes and perceptions.

		Religion (n=34)	Prejudice (n=195)	Personal experience (n=65)	F(2,291) (η^2)	
Attitudes	Total attitudes	2.82	2.99	3.07	4.79**	Personal experience
		(0.41)	(0.39)	(0.38)	(.032)	+ prejudice "/> Religion
	Closeness	2.92	3.13	3.27	5.27**	Personal experience "/>
		(0.53)	(0.51)	(0.48)	(.035)	Religion
	Emotions	3.20	3.35	3.38	1.12	
		(0.70)	(0.55)	(0.55)	(.008)	
	Cognitive	2.23	2.44	2.52	3.60*	Personal experience "/>
		(0.48)	(0.50)	(0.56)	(.024)	Religion
Perceptions / knowledge	Total perceptions	2.56	2.58	2.61	0.13	
		(0.44)	(0.50)	(0.49)	(.001)	
	Life skills	2.35	2.24	2.33	0.59	
		(0.58)	(0.69)	(0.68)	(.004)	
	Academic skills	2.78	2.93	2.83	1.00	
		(0.70)	(0.70)	(0.74)	(.007)	
	Social integration	2.62	2.66	2.72	0.45	
		(0.45)	(0.54)	(0.53)	(.003)	

*p<.05, **p<.01, ***p<.001

Table 3. Comparisons on attitudes and conceptions regarding persons with autism in the community according to main reason underlying attitudes and conceptions: religion, prejudice, personal experience

http://dx.doi.org/10.5772/54845

From table 3 it appears that religion was the least chosen by respondents (11.6%) as being the underlying factor affecting attitudes and conceptions regarding autism. Most (66.3%) respondents pointed at prejudice as being the most common factor underlying attitudes and conceptions, followed by the factor of personal experience (22/1%). Thus the first part of hypothesis 4 was not confirmed. The second part indicated that the factor of religion will be exhibited by the most negative attitudes and conceptions. Indeed, as seen from table 3, the mean scores obtained by those who pointed at religion as being the main factor, was the lowest, indicating negative attitudes and conceptions. Comparing the scores obtained on the three factors of religion, prejudice and personal experience the most positive attitudes and conceptions regarding persons with autism was a combination of personal experience coupled with prejudice (p < 01). On the subscale of willingness to be close to a person with autism personal experience had a significantly greater impact than religion (p<.01).

No differences were found regarding the relative impact of religion, prejudice or personal experience on perceptions regarding their ability to be included in the community.

We were further interested to find out whether any of the background variables of age, gender or having heard about autism before starting academic studies affected attitudes and perceptions.

		Yes (n=182)	No (n=126)	F(1,306) (η^2)
Attitudes	Total attitudes	3.05 (0.38)	2.91 (0.39)	10.27*** (.032)
	Closeness	3.21 (0.50)	3.03 (0.51)	9.31** (.030)
	Emotions	3.43 (0.52)	3.24 (0.60)	8.94** (.028)
	Cognitive	2.47 (0.50)	2.42 (0.54)	0.61 (.002)
Perceptions / knowledge	Total perceptions	2.62 (0.47)	2.50 (0.52)	4.67* (.015)
	Life skills	2.32 (0.62)	2.18 (0.71)	3.17 (.010)
	Academic skills	2.96 (0.65)	2.75 (0.76)	6.34* (.020)
	Social integration	2.69 (0.53)	2.61 (0.54)	1.79 (.006)

*p<.05, **p<.01, ***p<.001

Table 4. Comparison between respondents who heard about autism (N=182) and those who have not heard (N=126) about it on attitudes towards autism and conceptions on inclusion

No correlations were found between the background variables of age and gender however, respondents who had heard about autism before they started their academic studies – whether general of in special education, were more inclined to have positive attitudes and positive perceptions of their inclusion. A Multivariate analyses of variance was applied to all scores compared on having heard or not about autism.

Table 4 presents comparisons between the scores obtained by respondents who heard about autism (N=182) and those who have not heard (N=126) about it. Missing data – 13 respondents.

From table 4 it appears that having heard about autism resulted in a more positive attitude towards them ($p<.001$) though it did not affect a better knowledge regarding the syndrome of autism, and a more positive perception of their overall ability to be included ($p<.05$) in the community, especially their life skills abilities and their academic skill, but not their social skills.

5. Conclusion

The major conclusion of our study is that the issue of attitudes towards persons with autism and perceptions regarding their abilities to be included in the community are complex and require further research in order to find out the most effective way to induce positive changes in this area. The fact that in our study correlations were highly significant between positive attitudes towards persons with autism and positive perceptions regarding the possibility of their inclusion in the community, indicate that a change in attitudes from negative to positive should be followed by a change in the opinion that persons with autism can be included in the community. However our findings revealed that this was not necessarily the case.

Attending a course on autism, though it affected a general change in attitudes, from negative to more positive attitudes towards autism, it did not affect a change in in the perceptions regarding the capability of persons with autism to be independent and live in the community. So though a positive change occurred in the area of attitudes, it was not correlated with a similar positive change in one's acceptance of the fact that persons with autism can learn independent living skills, can acquire the academic skills necessary for a life in the community and can be socially included.

Apparently attending a course on autism turned out to be a necessary but not sufficient condition for a complete change of opinions regarding the possibility of persons with autism to live in the community. As the saying goes: 'old attitudes die hard'. Indeed incongruence between the different dimensions of attitudes, the emotional reaction, the behavioral aspect and the precision of knowledge regarding autism are not necessarily harmonious (Reiter and Bryen, 2010). Tackling the emotional dimension is highly important as it may pave the way for students to be ready to learn facts about autism that are not loaded with misconceptions and prejudices. Indeed, prejudice, more so than religion, was found to be the most significant force affecting attitudes and perceptions regarding autism. Another factor found to affect attitudes and conceptions was a personal experience with a person with autism. This finding suggests

http://dx.doi.org/10.5772/54845

that in order for an academic course on autism to succeed in changing negative conceptions it should be followed by an experiential process in which students should meet and interact with persons with autism and their families.

The findings of our research confirms what most publications regarding autism in the Arab world, such as Al-Shammari, (2006) point out, that religion is the major source of negative attitudes and perceptions regarding persons with autism. Also the report written by Estela Dimes (2012) indicates that religion can affect helplessness and passivity due to a belief that since having a child with autism is God's will, there is not much one can do about it. A person with autism is inflicted by a disability that gives him/her no chance of leading a 'normal' life.

However religion and faith can have two different and opposed outcomes as shown by the qualitative research conducted by Al-Shich with Druze mothers. In her study, some mothers said it gave them strength. Religion can be a source of strength, it can affect an attitude that it is God's way of testing me to be the best parent possible, and give me strength to provide my child with the best possible support for a life of quality.

In our study prejudice was found to be the most commonly cited factor underlying attitudes and conceptions regarding autism. Unlike religious faith, prejudice is one-sided. It is exhibited in negative attitudes and a rejection of persons with autism from the mainstream of society. The misconceptions that underlay prejudice are highly imbedded in our culture, starting from stories and legends told to toddlers and creating stereotypes followed by 'popular' derogatory and rejecting statements expressed by adults. Attitudes change calls for a comprehensive cultural reform and a change of outlook regarding disability in general and autism in particular. The findings that those among the respondents who heard about autism before starting their studies showed more positive attitudes towards them and the partial but significant effect of attending a course on autism are encouraging. Higher exposure of persons with autism and their families, more self-advocacy activities done by them, more courses provided in this area and better services, will hopefully contribute to a more positive attitude and more positive conceptions regarding their inclusion in the community.

Further research

Our study calls for more research in the areas of prejudice and attitude change regarding the Autistic Spectrum Disorders. Negative attitudes and lack of precise knowledge about the syndrome of autism is not unique to the Arab world. It is also not unique to the common 'person in the street', it affects service providers and professionals as well. In spite of the fact that there is a worldwide rise in the number of children diagnosed as having an autistic disorder, the medical knowledge about autism and its causes is still incomplete. Thus, there are a lot of unclear and unanswered questions regarding its causes, treatment, the best services needed to assist in their inclusion in mainstream society. One thing is though unquestionable, the rights of persons with autism for the best possible life with dignity and fulfillment (Reiter, 2008). A humanistic human rights philosophy should underlie not only our treatment and education of persons with autism but should also guide any future research in this area.

Author details

Karni-Vizer Nirit and Reiter Shunit*

*Address all correspondence to: shunitr@edu.haifa.ac.il

Faculty of Education, Department of Special Education, University of Haifa, Haifa, Israel

References

[1] Al HajM. ((1989). Social research on family life style among Arabs in Israel. *Journal of comparative family studies, 175194*, 20

[2] Al-shammari, Z. (2006). Special education teachers' attitudes toward autistic students in the autism school in the State of Kuwait. *Journal of Instructional Psychology, 33170178*

[3] Al-shich, M. (2012). Acceptance, coping and the perception of the future of the child with autism in the eyes of Druze mothers. MA Thesis. Haifa: Faculty of Education, University of Haifa.

[4] Devine, P. G. (1989). Stereotypes and prejudice: Their automatic and controlled components. *Journal of Personality and Social Psychology, 518*, 56

[5] Dimes, E. (2012). *Culture and Autism*. Feature Article, MinorityNurse.com

[6] Elbahnasawy, H, & Girgis, T. N., M. ((2011). Counseling for Mothers to Cope with their Autistic Children. *Journal of American Science, 7183192*1545-1003

[7] Flood, L. N, Bulgrin, A, & Morgan, B. L. (2012). Piecing together the puzzle: development of the Societal Attitudes towards Autism (SATA) scale. *Journal of Research in Special educational Needs*. Available online in: Doi:j/x, 1471-3802.

[8] Kandel, J, Morad, M, Vardi, G, Press, J, & Merrick, J. (2004). The Arab community in Israel coping with intellectual and developmental disability. *Scientific World Journal, 4*, 324–332.

[9] Karni, N, Reiter, S, & Bryen, N. D. (2011). Israeli Arab teachers' attitudes on inclusion of students with disabilities. *The British Journal of Developmental Disabilities, 113121130*

[10] Mar, i. S. ((1978). Arab education in Israel. Syracuse, USA: Syracuse University Press.

[11] Mar, i, & Karayanni, S. M. ((1982). Creativity in Arab culture: Two decades of research. *Journal of Creative Behavior, 16227238*

[12] Mostafa, A. (2011). *Addressing autism in the Arab world*. Available online in: 2 November in: doi:10.1038/nmiddleeast.2011.

[13] Odeh, M. (2007). Pathogenesis of hepatic encephalopathy: the tumour necrosis factor-alpha theory. *European Journal of Clinical Investigation, 37(4),* 291-304. April 2007. PMID: 17373965 [PubMed- indexed for MEDLINE]

[14] Reiter, S, Mar, i, & Rosenberg, S. Y. ((1986). Parental attitudes toward the developmentally disabled among Arab communities in Israel: A cross-cultural study. *International Journal of Rehabilitation Research, 9355362*

[15] Reiter, S. (1999). *Society and disability: An international perspective on social policy.* Haifa: AHVA publishers.

[16] Reiter, S. (2008). *Disability from a Humanistic perspective.* New York: Nova Biomedical Books.

[17] Reiter, S, & Bryen, D. N. (2010). Attitudinal barriers to rehabilitation. *International Encyclopedia of Rehabilitation,* JH Stone, M. Blouin, editors. Available online in: http// cirrie.buffalo.edu/encyclopedia/article.

[18] Rosenbaum, P, Armstrong, R. W, & King, S. M. (1986). Children's attitudes toward disabled peers: a self-report measure. *Journal of Pediatric Psychology, 11517530*

[19] Sandler-loeff, A, & Shahak, Y. (2006). *People with disabilities in Arab society in Israel: An opportunity for social change*. Jerusalem: JDC Israel, The Unit for Disabilities and Rehabilitation.

[20] The American Heritage® Dictionary of the English LanguageFourth Edition copyright ©(2000). by Houghton Mifflin Company. Updated in 2000.

[21] Tirosh, E, Schanin, M, & Reiter, S. (1997). Children's attitudes toward peers with disabilities: The Israeli perspective. *Developmental medicine & Child Neurology, 39811814*

[22] Vardi-saliternik, R, Friedlander, Y, & Cohen, T. (2002). Consanguinity in a population sample of Israeli Muslim Arabs, Christian Arabs and Druze. Ann Hum Biol. *29442231*

Section 3

Brain and Autism

Chapter 3

The Role of Cortical Modularity in Tactile Information Processing: An Approach to Measuring Information Processing Deficits in Autism

Eric Francisco, Oleg Favorov and Mark Tommerdahl

Additional information is available at the end of the chapter

http://dx.doi.org/10.5772/54801

1. Introduction

Autism is a pervasive developmental disorder that is manifested in a number of neurological alterations. Although there is a large spectrum of behavioral excesses that includes a diverse number of traits, such as repetitive behaviors and/or sensory hyper-responsiveness, many of the neurological problems could be attributed to underlying anatomical and physiological fundamentals that demonstrate significant diversity within this spectrum and make the phenotypic description of the disorder distinctly different from that exhibited by normal physiology. Characterization of neurological features – such as cortical modularity – could lead to a better understanding of the neurophysiological fundamentals of autism. Recently, we have been developing sensory-based diagnostic protocols based on neurophysiological principles that have been elucidated in animal studies conducted both in our laboratories and those of others. One question that we have pursued in our animal studies has been the fundamental role(s) of the cortical minicolumn and macrocolumn in tactile information processing. We have developed experimental models for determining cortical correlates of perception that relate cortical activity patterns in somatosensory cortex (at high resolution in squirrel monkey studies) to measures of human perception. The minicolumnar and macrocolumnar organization of the cerebral cortex is dynamic and interactive, and the patterns of activity that are generated with stimulus-driven activity in SI cortex have been shown to be modular in nature. This determination of modularity is derived from a self-organizing process that takes place via dynamic interactions between minicolumns and columns in the cortex both during and after development. If developmental processes malfunction, then cortical organization suffers at a number of scales. Findings by Casanova and colleagues have elegantly demonstrated in post-mortem histological experiments that minicolumn organization in autism is severely compromised, as there are approximately 30% more minicolumns in the same cortical space as is normally found [1]. The increase in minicolumn density, and particularly the decrease in neuropil between the minicolumns (because they are now much more densely packed), led us to make a number of predictions about alterations in perceptual metrics that would occur in individuals with autism. In this paper, the neurophysiological basis of three such perceptual metrics (previously reported) is discussed.

2. Cortical modularity and spatial localization

In 1978 Mountcastle [2] hypothesized that the smallest functional unit of neocortical organization, the "minicolumn", is a radial cord of cells about 30-50μm in diameter, and that sensory stimuli activate local groupings of minicolumns (called "macrocolumns"). This hypothesis subsequently received support from multiple lines of experimental evidence and led to its substantial elaboration. Structurally, minicolumns are attributable to the radially-oriented cords of neuronal cell bodies that are evident in Nissl-stained sections of the cerebral cortex and it is probable that they are related to ontogenetic columns [3] and to the radially-oriented modules defined by the clustering of the apical dendrites of pyramidal neurons [4]. Among the various elements of neocortical microarchitecture, *spiny-stellate* cells and *double-bouquet* cells [5-7] are most directly relevant to Mountcastle's concept of the minicolumn. Spiny-stellates are excitatory intrinsic cells that are especially prominent in layer 4 of primary sensory cortex. They are the major recipients of thalamocortical connections and, in turn, they distribute afferent input radially to cells in other layers. Double-bouquet cells are GABAergic cells whose somas and dendritic trees are confined to the superficial layers, and because the double-bouquet cells are more likely to inhibit cells in adjacent minicolumns rather than in their own, they offer a mechanism by which a minicolumn can inhibit its immediate neighbors.

Some insights into the role of the minicolumn in sensory information processing have been revealed through neurophysiological experimentation. Receptive field mapping studies by Favorov and colleagues [8] determined that there are abrupt shifts between receptive field centers as stimuli shift from one skin site to another. In other words, Favorov's receptive field work predicted that a perceptible but subtle shift of stimulus position would not necessarily engage a different pattern of macrocolumnar activity. Rather, the pattern of minicolumnar activity *within a macrocolumn* would be different with a shift in stimulus position up to a point. At the point at which the stimulus position crosses a boundary, the stimulus will engage a new macrocolumn and an entirely different minicolumnar pattern of response will be evoked by the stimulus. Figure 1 summarizes minicolumnar RF organization in the somatosensory cortex. Note that as an electrode penetration moves tangentially across a field of cortical macrocolumns (note locations of penetrations 1-30), the receptive field center (indicated on the digit tips to the right of the cortical field) moves a significant distance only after crossing a macrocolumnar border. While within the macrocolumn, the receptive field centers remain relatively closely spaced. It is also of note that the receptive field properties are constrained within the radial dimension; that is, if the electrode is moved along the radial dimension (note the penetrations denoted a-g), the receptive field center does not shift and receptive field proper-

http://dx.doi.org/10.5772/54801

ties will be very similar. As the description above is over-simplified, it should be noted that there is a great diversity of receptive field properties between neighboring minicolumns, and a stimulus that effectively activates one minicolumn will often be ineffective at activating that minicolumn's nearest neighbor [9-10]

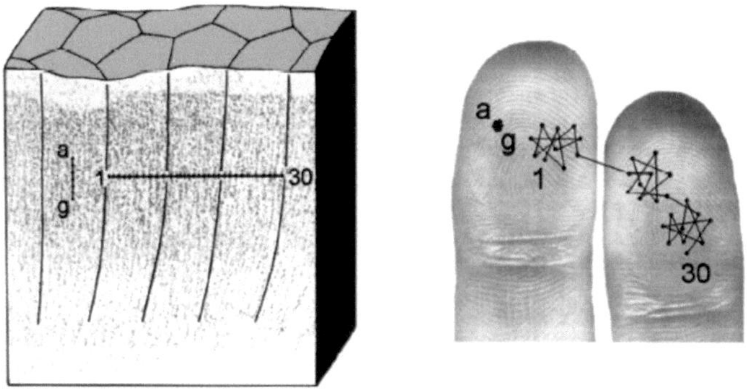

Figure 1. Summary of minicolumnar RF organization in SI somatosensory cortex. Left: Drawing of cross-section of Nissl-stained cortical tissue showing darkly-stained cell bodies organized in radially oriented cords, interpreted as minicolumns. Filled circles labeled *a–g*—sequence of neurons located within a single minicolumn; *1–30*—sequence of neurons located in series of adjacent minicolumns. Right: Sequences of RF centers (connected dots) mapped by neuron sequences *a–g* and *1–30*. Note that RF centers for SI neurons that occupy the same minicolumn stay close together, whereas the RF centers for pairs of neurons located in neighboring minicolumns shift back and forth over large distances, and occupy totally non-overlapping skin regions when the pair of neurons occupies different SI macrocolumns. Based on [8-10].

The findings and predictions by Favorov et al were later confirmed with additional data that was obtained via optical intrinsic signal imaging (for description of technique, see [11]). In this study, responses evoked by vibrotactile stimuli delivered to different positions on the skin (which differed by only the width of the 2mm probe tip) showed a subtle variation within the macrocolumnar pattern within a range of stimulus positions (analyzed with the methods described in [12]), but the global pattern did not shift until a new group of minicolumns (or macrocolumns) was stimulated. Figure 2 summarizes the results of one such imaging experiment in which the macrocolumnar pattern of cortical response does not significantly alter with a small shift in stimulus position until a border is crossed. Additional features of these minicolumnar patterns of activity that have been characterized are that they are stimulus magnitude- and duration-dependent [12-14]. For example, increasing the stimulus duration leads to more distinct and well-defined minicolumnar patterns of cortical activity. Additionally, the spectrum of the spatial profile of this activity evoked by the active minicolumns robustly and significantly shifts to lower frequencies, and the spectral shifts that have been observed are consistent with the concept of increased GABA mediated lateral inhibition between minicolumns [12]. Perceptually, these changes in minicolumnar activity patterns with stimulus duration could parallel the increases in sensory perception that have been observed with longer stimulus durations [15].

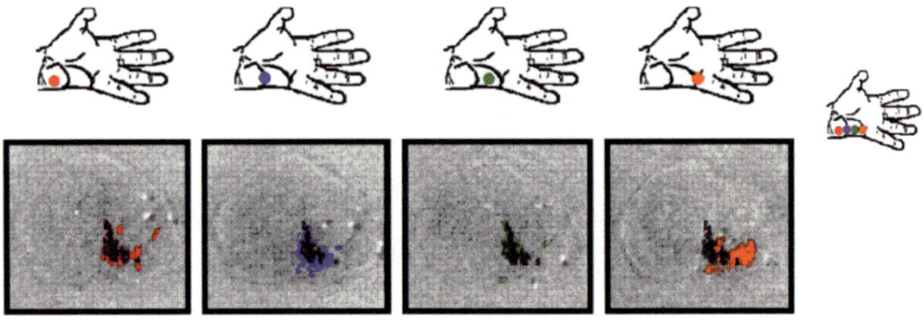

Figure 2. Summary of optical intrinsic signal evoked responses in SI cortex from four adjacent stimulus positions. Difference in stimulus position was equal to probe diameter (2 mm). Note the abrupt shift between the responses evoked from the stimulus placed at the first three positions vs. that of the fourth. Modified from [13].

Casanova and colleagues have demonstrated that there is a substantial increase in minicolumn density in the parietal cortex of individuals with autism [1,16]. This increase in minicolumn density results in a disproportionately large number of minicolumns becoming packed into the same cortical space and also results in a decrease in the neuropil between minicolumns. Thus, although there are now a higher density of minicolumns, there is less room for the GABA mediated lateral inhibitory connections between the minicolumns that are necessary for shaping the within-macrocolumn response that has been observed with repetitive stimulus duration [12-14, 17-20]. This alteration in basic cortical microarchitecture would then predictably contribute to an individual's sensory perception in a couple of ways. First, the increase in minicolumn density should afford an individual with autism an advantage in some sensory tasks, such as spatial localization, in which the percept would be improved. However, below baseline GABA mediated lateral inhibition between minicolumns would mean that increasing the duration of a stimulus would not increase the resolution or distinction of the within macrocolumn pattern of minicolumnar activity to the same degree, and thus, perception would not be improved. With this hypothesis of the minicolumn's role in spatial localization in mind, we designed an experiment to evaluate the differences between the spatial localization ability of neurotypical controls and subjects with autism [21]. In the study, a subject's ability to distinguish between two points on the skin (on the hand dorsum) was determined with two different stimulus durations – 500 msec and 5 sec (full description of the method in [21, 22]). Results from that study are summarized in Figure 3. Although individuals with autism outperformed controls in the shorter stimulus duration task, they did not demonstrate the nearly two-fold improvement that the controls did when the stimulus duration was extended. Thus, in the case of spatial localization, it appears that alterations in sensory percept could be accounted for by the changes that have been observed in cortical minicolumn architecture.

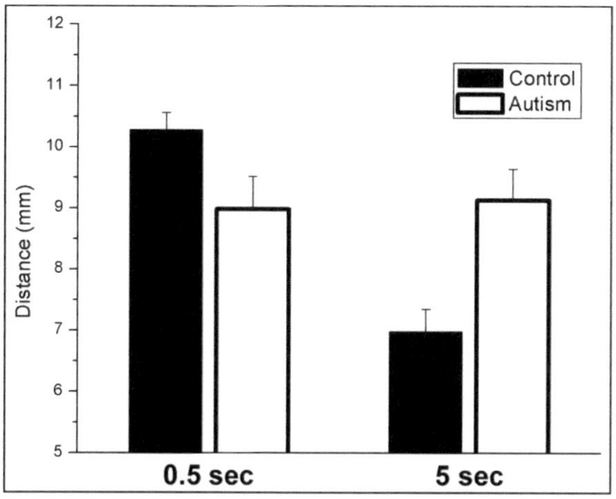

Figure 3. Spatial localization under two conditions of adapting stimulus duration for adults with autism versus neurotypical controls. Data displayed from the control subjects (previously reported [22]) contrasts markedly from the data obtained from observations of subjects with autism. Note that subjects with autism, although they clearly outperformed the controls in the 0.5 sec adapting condition, did not improve with the 5.0 sec adapting condition. Modified from [21]

The difference that was observed in short vs. long stimulus duration in the above-described spatial localization experiment led us to examine more directly the relationship between our previous adaptation animal studies and the role that adaptation – or conditioning stimulation – plays in sensory information processing in autism. It has been well established that conditioning stimulation – or prolonged pre-exposure to sensory stimulation – significantly modifies discriminative capacity and alters the ability of both peripheral and CNS neurons to process sensory information. Less widely appreciated is the fact that primary sensory cortical mechanisms undergo transient and significant alterations in response to repetitive sensory stimulation. Investigation of the dynamic cortical responses evoked by repetitive stimulation has been an ongoing line of research in our laboratory. One of the focal points has been the spatio-temporal patterns of response in the somatosensory cortex evoked by skin stimulation and how these patterns influence the cortical response to subsequent stimuli. For example, the observations of a number of studies have demonstrated that the spatially distributed pattern of activity evoked in SI cortex by cutaneous flutter stimulation exhibits a prominent time-dependency [11, 23, 24]. Specifically, changing the stimulus duration from 500 msec to 5 sec (such as was done in the spatial localization task described above) would result in two distinctly different patterns of response in SI cortex. Figure 4 compares the profiles of two SI cortical responses evoked by vibrotactile stimuli that differed only in duration (Note that the profile is a radial histogram of OIS images generated by plotting the cortical activity evoked by the stimulus as a function of the distance from the center of the region in SI that is maximally activated by the stimulus; [23, 24]). With the 500 msec stimulus, the entire response profile is above-background. However, with the longer duration 5 sec stimulus, a suppressive or inhibitory region surrounds the maximally activated region. This region of inhibitory influence – which persists for several seconds – would interfere with the SI response to a stimulus applied concurrently or subsequently to skin regions in near proximity represented by neurons in that region of SI. Thus, in the case of the above-described spatial localization task, longer stimulus durations would be expected to improve performance. Since the presence of a center-surround in stimulus evoked cortical activity is commonly recognized as a function of GABA mediated pericolumnar lateral inhibition [25, 26], and a number of researchers have described GABA deficiency as being consistent with autism [27-31], we concluded that the lack of improvement with increasing stimulus duration in autism subjects in the spatial localization task could be due to a deficiency in GABA mediated neurotransmission.

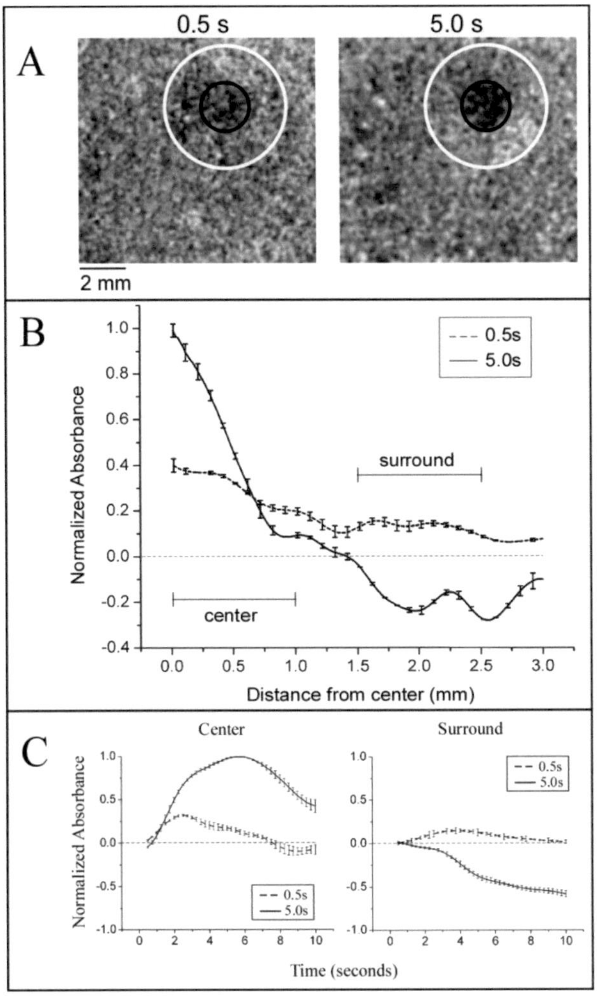

Figure 4. Radial histograms of SI cortical activity averaged across subjects (squirrel monkeys, $n = 5$). Cortical activity is measured in terms of light absorbance (increased light absorbance can be correlated to increased cortical activity, for review, see [11]). Data from the center of the plot corresponds to the maximally responding SI cortical territory to the 5 sec stimulus condition. Note that at the 0.5 sec stimulus duration, there is no below-background activity. Modified from [23].

The improvements that are normally observed with extended stimulus durations could be attributed to stimulus-evoked inhibition that surrounds areas of excitation. Single unit studies and imaging studies using voltage-sensitive dyes likewise have shown that excitation in the responding neuronal population is accompanied by the development of a surrounding field of inhibition [32-35]. Similarly, imaging studies that have used the OIS have shown that prolonged stimulation of a discrete skin site not only is associated with increased absorbance within the SI region representing the stimulated skin site, but also with decreases in absorbance in surrounding regions [23, 36-38]. Regions of decreased absorbance (increased reflectance) such as that described in Figure 4 are widely believed to be indicative of decreases in neuronal spike discharge activity [39-41], possibly resulting from stimulus-evoked inhibition at these locations. Thus, there is a great deal of evidence that the suppressed or below-background activity observed suggests that stimulus-evoked inhibition is responsible for the improvements in performance that are normally observed with repetitive stimulation. However, it appears that in the case of autism, there is sufficient evidence to speculate that the normal center-surround relationship in cortical patterns of activity does not fully develop.

3. Cortical modularity and adaptation

In addition to changes in spatial contrast, as described above, repetitive stimulation also results in temporally defined changes of cortical activity, the most prominent of which is a reduction in cortical response with extended stimulus duration. At the single cell level, both visual and somatosensory cortical pyramidal neurons undergo prominent use-dependent modifications of their receptive fields and response properties with repetitive stimulation. These modifications can attain full development within a few tens of milliseconds of stimulus onset, and can disappear within seconds after the stimulus ends (visual cortical neurons: [42-53]) alternatively – for review of short term cortical neuron dynamics in visual cortex, see [53, 54]; for review of short-term primary somatosensory cortical neuron dynamics see [15, 55].

Optical imaging studies have also characterized the short-term dynamics of the population-level response of squirrel monkey contralateral primary somatosensory (SI) cortex using different amplitudes and durations of vibrotactile stimulation [11, 12, 23, 24, 56]. The results of these optical intrinsic signal (OIS) imaging studies demonstrated a strong correlation between the amplitude of 25 Hz vibrotactile (flutter) skin stimulation and the response magnitude evoked in SI. In addition to the systematic changes in the spatial pattern of response in SI that correlated with increases in the amplitude and the duration of the stimulus, increasing the stimulus duration led to differences not only in the peak magnitude of the evoked cortical response, but also in the relative rates of rise and decay of the magnitude of the evoked intrinsic signal. These differences in the rates of rise and decay could impact the refractory period following a stimulus during which the magnitude of the response to a subsequent stimulus is diminished [57].

In order to assess the impact that adaptation has on perception, experiments were designed to directly measure the change in amplitude discrimination capacity that occurs with prior stimulus exposure (or prior conditioning stimuli). The studies demonstrated that a subject's ability to discriminate between two simultaneously delivered vibrotactile stimuli – differing only in amplitude and location – was very robust and repeatable across a large number of (healthy) subjects but was very sensitive to varying conditions of pre-exposure to sensory stimuli [58]. Changing the duration of the conditioning stimulus delivered to one of the two sites before the amplitude discrimination task significantly altered a subject's ability to determine the actual difference between the two stimuli. One significant finding of that study was that specific durations of conditioning stimuli altered the subject's amplitude discriminative capacity in a predictive and quantifiable fashion (see Figure 5).

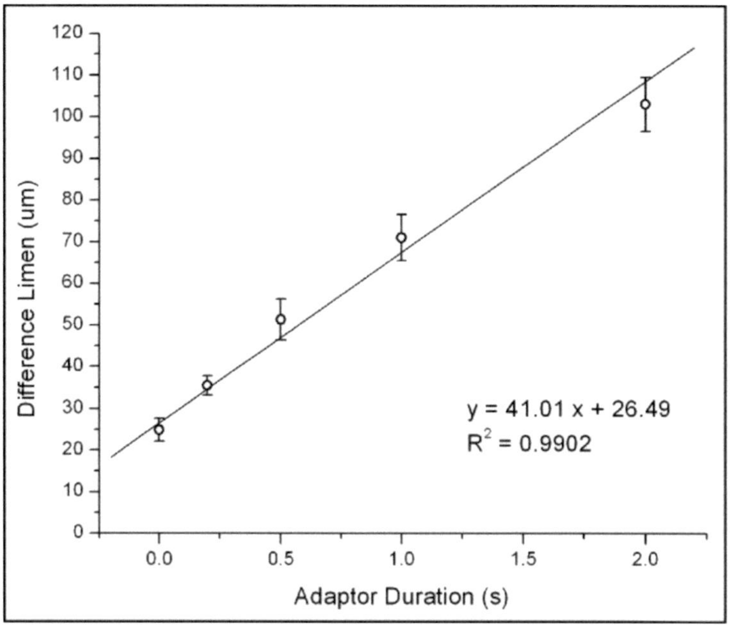

Figure 5. Comparison of amplitude difference threshold (with s.e. bars) to different conditions of adaptation. The test and standard stimuli were preceded by an adapting stimulus at the site of the test stimulus (ranging from 0.2 to 2 sec in duration). Note that single site adapting stimulation leads to a progressive and systematic decrease in performance with increasing adaptor duration [58].

This finding indicated that the method could be viewed as a reliable indicator of the influence of adapting stimuli on cortical response, as changes in peripheral response are not mediated at these short stimulus durations (for discussion, see [58]).

Conditioning stimuli did not have as pronounced an impact on the amplitude discriminative capability of subjects with autism as it did with the control group [60]. In Figure 6, results obtained using identical methods from subjects with autism and controls are compared. Note that adaptation (i.e., a 1 sec conditioning stimulus at one stimulus site prior to the amplitude discrimination task) resulted in the control subjects performing significantly worse than they did in the absence of adaptation. However, in the case of the autism subjects, the impact of prior history of stimulation was not as significant, and the amplitude discrimination metric was not impacted to the same degree as it was in the controls. Thus, the ineffectiveness of a conditioning stimulus in this study repeated the findings of the spatial localization studies, in that adapting stimuli had little or diminished effect – positive or negative – on the sensory discriminative performance of individuals with autism.

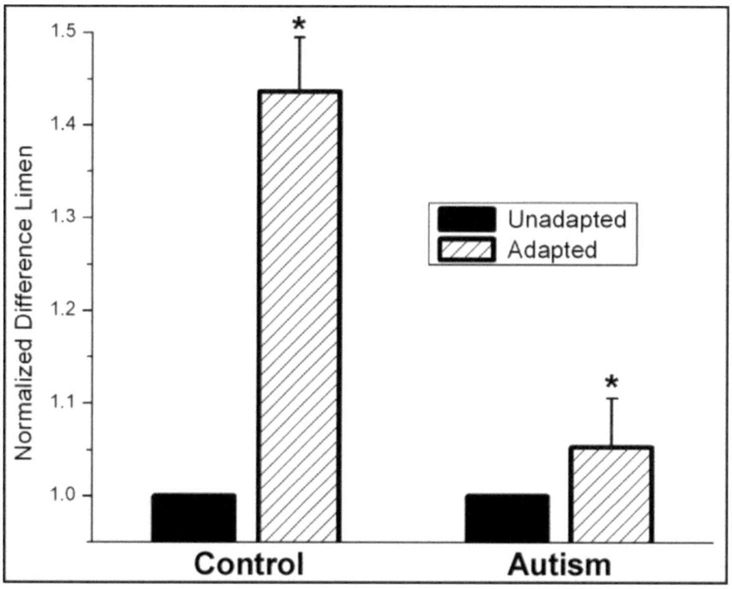

Figure 6. Comparison of difference limen (with s.e. bars) normalized to the unadapted condition. Note that for both the control and the autism group, 1 sec adaptation resulted in an elevated difference limen (* ANOVA; $p < 0.01$). The control group showed a greater impairment with adaptation (~45%) than the autism group (~5%) [59].

4. Cortical modularity and synchronization

There are a number of autism studies that have described Parkinsonian-like motor characteristics and/or postural control problems, which could be attributed to deficits of the basal ganglia portion of the frontostriatal system [60, 61]. These deficits in sensorimotor control could be derived, in part, from the role that the frontostriatal system plays in an individual's timing perception as well as the coordination that is required between cortical regions during sensorimotor tasks. Timing perception, which can be measured with some relatively simple temporal discriminative measures (such as TOJ: temporal order judgment and TDT: temporal discriminative threshold) – is most often accounted for by the frontostriatal system largely as a result of these measures being sensitive to lesions to the supplementary motor area (SMA), posterior parietal cortex, and basal ganglia [62, 63]. Also because of the fact that above-average TOJ thresholds occur in subjects with known damage to these same cortical areas (dyslexia [64], dystonia [65-67] and Parkinson's disease [68]). Most recently, it was found that individuals with autism also have below-average timing perception capacity [69]. This timing deficit could be accounted for by differences in a number of structures, particularly in the frontostriatal system, that have been implicated in autism (e.g., basal ganglia [70-76]; caudate nucleus [70]; thalamus [77, 78]; and impaired white matter connectivity in the frontal lobe [79]).

In addition to the role that the frontostriatal system may have in the perceptual timing deficits of autism, the role of synchronization (or lack of synchronization) in autism has gained a certain degree of prominent attention. Uhlhaas and Singer [80] recently reviewed the experimental evidence that suggests that functional connectivity is reduced in autism, primarily based on fMRI studies [81-86] that examine the coordinated activity between different areas of the cerebral cortex. A few studies, using MEG and EEG, have found gamma oscillations, which are considered to be important in the process of coordinating cortical activity, to be below normal in subjects with autism [87, 88]. From the perspective of cortical modularity at both the minicolumnar and macrocolumnar scales, synchronization at the local cortical level should also be impacted. Casanova and colleagues have suggested that the aberrant minicolumnar structure that they have found in autism could result in the disruption of the inhibitory architecture [16] that is required for normal function in local neural circuitry. Disruption of functional connectivity at the local minicolumnar level could be responsible for, or strongly correlated with, the dysfunctional connectivity that has been observed across large-scale cortical areas.

There is a rapidly growing appreciation in neurobiological research of the important contributions to sensorimotor function of coordinated across-neuron patterns of spike discharge activity within the neocortical areas activated by sensory stimuli (for comprehensive review see [89]). In particular, stimulus-induced, time-dependent (dynamic) across-neuron synchronization of action potential discharge and the associated oscillatory modulation of spike firing are common and prominent properties of neocortical networks devoted to the processing of sensory information. The tendency of sensory neocortical networks to generate synchronized oscillations in response to stimulation has raised the possibility that synchronization may play a prominent role in some aspects of sensory perception. We examined whether or not synchronization could impact the topography of temporal perception [90]. The goal of the study was to elucidate the impact of stimulus-driven synchronization on adjacent cortical ensembles and the spatio-temporal integration of information that results from those ensembles being temporally linked or bound by a common synchronizing input. More specifically, we demonstrated that temporal order judgment (TOJ – a measure obtained from determining the minimal inter-stimulus interval necessary for a subject to detect the temporal order of two sequentially delivered peripheral stimuli) and temporal discrimination threshold (TDT) in neurotypical subjects were significantly impacted when two synchronized (but low amplitude) vibrotactile stimuli were delivered concurrently to the dual test stimulus sites. The conclusion of that study was that the stimulus-driven linkage between topographically adjacent sites resulted in an increase in TOJ threshold and TDT (or worse performance), most likely because these cortically adjacent or near-adjacent regions were being driven with a simultaneous and identical sinusoidal pair of tactile stimuli which contributed to a loss in spatio-temporal contrast [90].

http://dx.doi.org/10.5772/54801

A subsequent question that was then addressed was whether or not individuals with autism experience a decrease in timing perception (as measured by TOJ and TDT) if the same concurrent synchronizing stimuli were delivered during the TOJ/TDT tests. If neurologically compromised individuals – such as those with autism – have distinct systemic cortical deficits, and these deficits extend to local neuronal circuitry connectivity, then the abnormal functional connectivity between adjacent and/or near adjacent cortical ensembles would hypothetically decrease the effect that stimulus-driven synchronization has on the TOJ or TDT task (i.e., performance on the task would not degrade). Comparisons of the control vs. autism results (previously reported in [69, 90]) are shown in Figure 7. Note that with concurrent stimulation, individuals with autism do not suffer the same decrease in sensory discriminative performance that controls do. In other words, the functional linkage in controls that becomes rapidly established, due to local synchronization effects, appears to perceptually bind the two stimulus sites (in this case, digits two and three) to an extent that it becomes more difficult to identify the temporal order between the two sites. Thus, as in the case of adaptive responses, it appears that there is a loss of an ability to integrate both spatial and temporal information in autism.

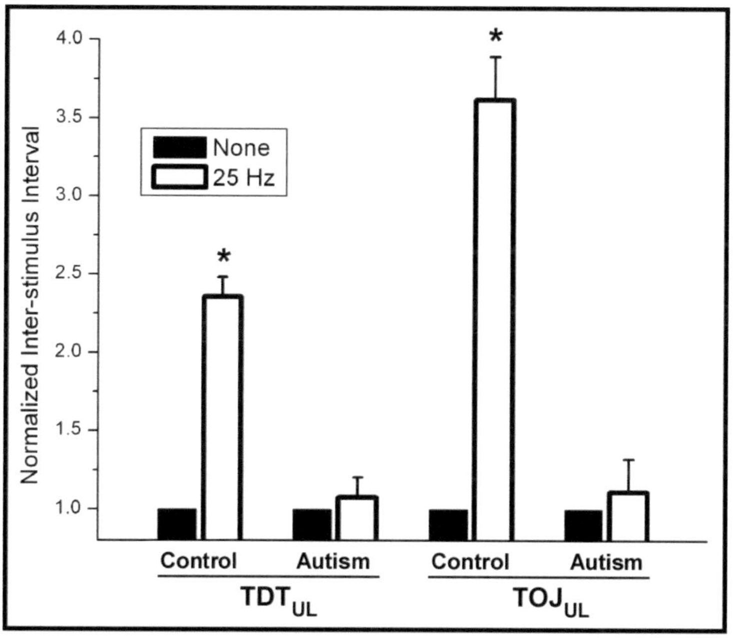

Figure 7. TDT and TOJ performance metrics obtained in the presence and absence of 25 Hz conditioning stimulation. The 25 Hz conditioning stimulus significantly impaired TDT by ~240% ($p < 0.01$) and TOJ by ~360% ($p < 0.01$) for the control group, whereas the autism group showed no significant change for either measure [69].

What could account for the reduction in TOJ performance in Typically Developing (TD) controls? Functional connectivity between neighboring cortical regions normally leads to a reduction in TOJ performance in healthy controls with the introduction of the synchronized conditioning stimuli, and this is predicted by recordings from *in vivo* animal studies. Consider the results displayed in Figure 8. Extracellular recordings were obtained from SI cortical regions corresponding to D2 and D3 in the squirrel monkey. When a vibrotactile pulse was delivered to D2, a significant above background response was evoked at D2 (top left quadrant) but not at the D3 representation (top right quadrant). However, when sub-threshold synchronized sinusoidal stimuli were delivered to both digits prior to the pulse (bottom half of Figure 8), the pulse at D2 evokes a response at both the D2 and D3 representations (note absence of evoked activity before zero msec during subthreshold stimulation).

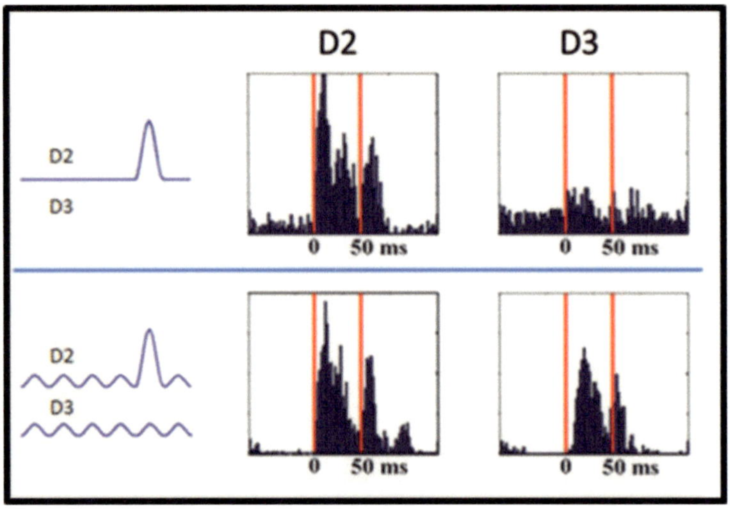

Figure 8. Extracellular recordings obtained from SI cortical regions corresponding to D2 and D3 in the squirrel monkey. When a vibrotactile pulse was delivered, a significant above background response was evoked at D2 but not at the D3 representation. When sub-threshold synchronized sinusoidal stimuli were delivered to both digits prior to the pulse, the pulse at D2 evokes a response at both the D2 and D3 representations.

From this type of data, we hypothesized that this response was the result of functional connectivity between adjacent and/or near adjacent cortical ensembles. In other words, the conditioning stimuli delivered prior to the TOJ task engaged the cortical ensembles in the D2 and D3 cortical representations to be in concert, and delivery of a simple stimulus to one digit (D2) resulted in a near simultaneous response at the representation of another digit (D3). Thus, it would be predicted that delivery of synchronized conditioning stimuli would impact the topography of temporal perception [90]. However, individuals with autism do not suffer the same decrease in sensory discriminative performance that neurotypical controls do. In other words, the functional linkage that becomes rapidly established in TD individuals to local synchronization effects. appears to perceptually bind the two stimulus sites does not occur in autism [69]. Thus, an extrapolation of this is that, utilizing measures impacted by stimulus driven synchronization, there is significant hypo-connectivity in autism at the level of local cortical ensembles.

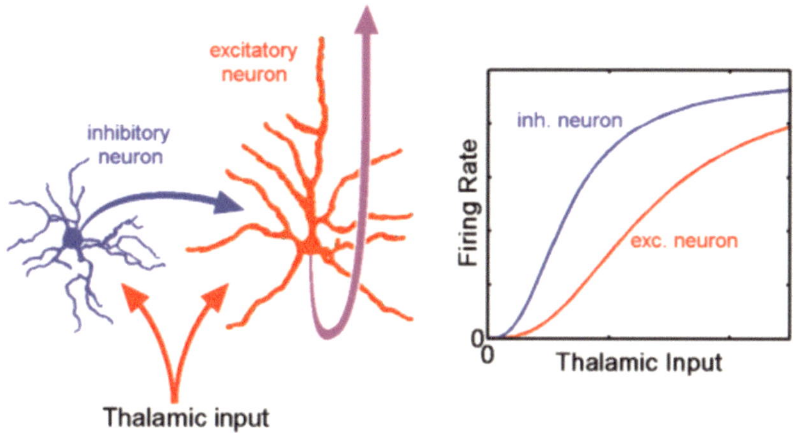

Figure 9. Visual representation of feed forward inhibition

Interim Summary: In the previous 3 sections, we described sensory-based diagnostic protocols that were based on neurophysiological principles that have been elucidated in animal studies. In the following sections, we describe additional protocols based on hypotheses that have not yet been tested in *in vivo* animal models, and it is anticipated that these protocols will add further insight into differences in fundamental mechanisms of information processing between TD and Autism Spectrum Disorder (ASD) individuals.

5. The role of sub-threshold stimulus-evoked inhibition: feed-forward inhibition and the role of within-column connectivity

A major well-documented feature of cortical functional organization is the presence of prominent feed-forward inhibition in the input layer 4 (see Figure 9). Local layer 4 inhibitory cells receive direct thalamocortical input and in turn suppress responses of neighboring layer 4 excitatory cells to their thalamocortical drive, thereby sharpening their RF properties [91-96]. These inhibitory cells are more responsive to weak (near-threshold) afferent drive than are the excitatory layer 4 cells and thus they *raise* the threshold at which excitatory layer 4 cells begin to respond to peripheral stimuli. Sensory testing of stimulus detection threshold is particularly well-suited for probing feed-forward inhibition, considering that stimuli just below the detection threshold will be too weak to vigorously engage other layer 4 mechanisms besides thalamocortical excitation and feed-forward inhibition (such as lateral excitation, recurrent or feedback inhibition, or activity-driven adaptation).

Tactile thresholds were collected in two distinct manners. The "static thresholds were measured using a 20-trial Two Alternative Forced Choice (2AFC) Tracking protocol. During each trial a 25 Hz vibrotactile test stimulus (lasts 500 ms) was delivered to either D2 or D3; the stimulus location was randomly selected on a trial-by-trial basis. Following each vibrotactile stimulus, the subject was prompted to select the skin site (D2 vs. D3) that perceived the stimulation. After a 5sec delay – based on subject response – the stimulation was repeated until the completion of the 20 trials. The stimulus amplitude was started at 15μm and was modified based on the subject's response in the preceding trial. During the "dynamic" threshold detection, a 25 Hz vibrotactile stimulus was delivered to either D2 or D3 (the stimulus location was randomly selected on a trial-by-trial basis). The amplitude of the stimulus was initiated from zero and increased in steps of 2 μm/s. The subject was instructed to indicate the skin site that received the stimulus as soon as the vibration was detected. Multiple trials were conducted with a random delay between trials and the results from those trials were averaged for each subject. For a complete experimental explanation, see [97-99].

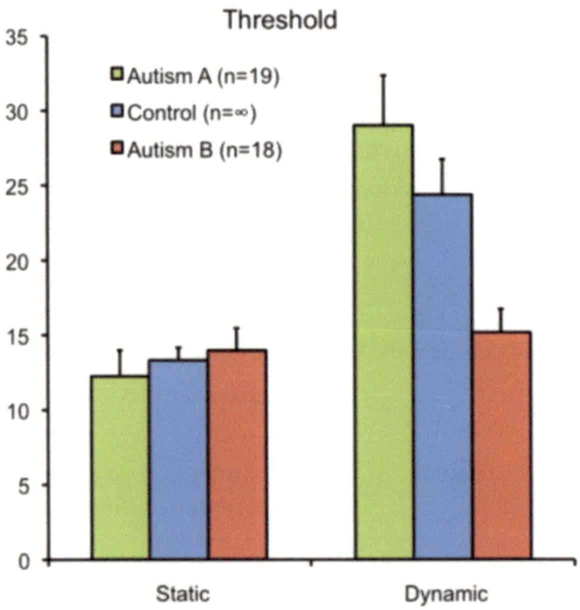

Figure 10. Static Thresholds collected using a 2AFC Protocol are compared to Dynamic Thresholds using a ramping amplitude protocol.

In our comparative study of typically developing vs. autism individuals, we found that subjects with autism exhibit significantly greater diversity in their detection thresholds on fingertips than control subjects, with two groups emerging (designated as Group A and Group B). Based on cluster analysis of several measures, the data that we have obtained thus far strongly suggests two distinct clusters within the spectrum. Group B autism individuals have dynamic thresholds lower than controls (thus suggesting reduced feed-forward inhibition) and group A autism individuals have dynamic thresholds higher than controls (thus suggest-

http://dx.doi.org/10.5772/54801

ing enhanced feed-forward inhibition). Inhibitory neurogliaform cells in layer 4 use both $GABA_A$ and $GABA_B$ receptor-mediated inhibitory synaptic transmission ([100]; in other inhibitory cell classes, $GABA_B$ receptors are located in the presynaptic membrane and used for autocontrol). $GABA_B$-mediated inhibition develops and lasts much longer than $GABA_A$-mediated inhibition. We believe we detect the $GABA_B$ component of feed-forward inhibition in our new "dynamic:" variant of the basic ("static") detection threshold test, in which we deliver vibrotactile stimuli of gradually increasing amplitude (starting at zero and growing at a rate of 2 μm/s) until the subject detects the vibration. Interestingly, this time-extended mode of stimulus delivery prominently elevates the detection threshold (compare "static" and "dynamic" plots in Figure 10), presumably by fully activating slow $GABA_B$ inhibition in addition to fast $GABA_A$ inhibition. Again we find that autism subjects exhibit greater diversity on this test than controls: group A autism individuals have static thresholds below controls, but dynamic thresholds above controls (suggesting reduced $GABA_A$ inhibition, but elevated $GABA_B$ inhibition), while group B autism have the opposite relations. Thus, if alteration of GABAa vs. GABAb inhibition influences the impact of subthreshold mediated activation, then the two aforementioned autism populations should, if treated pharmacologically, respond differently to a GABAb agonist, such as baclofen. If this is the case, then a simple measure such as that described above could predict whether or not this particular treatment would be effective.

6. Temporal integration: Rate dependent modulation of vibrotactile stimuli

The difference that we observed with static vs. dynamic thresholds encouraged us to explore the impact that changing the rate of amplitude modulation would have on sensory percept performance at supra-threshold levels. In the dynamic threshold task, an amplitude modulation rate of 2 μm/s was used to deliver a *subthreshold* stimulus. Delivering higher rates of amplitude modulation at above threshold values yields very different results. In the data presented below, a subject's ability to match two stimuli was assessed at nine different rates of amplitude modulation (one stimulus was held at steady state values, the other was increased until it was perceived as a match; [99]). At the lowest rates, subjects performed comparably to the dynamic threshold task (Autism group A performed worse than controls and Autism B, though not significantly). With an increasing rate of modulation, from 1.25 to 10 μm/s, the Autism A group demonstrated a decreasing Difference Limen (DL). As the rate increased above 15 μm/s ond, this group began performing significantly worse in that the test stimuli was increased well beyond the value of the standard stimulus (resulting in the "negative" DL). It appears that this group was unable to temporally integrate information from the stimuli, and future in vivo studies will examine the role that different neurotransmitter systems play in such temporal integration.

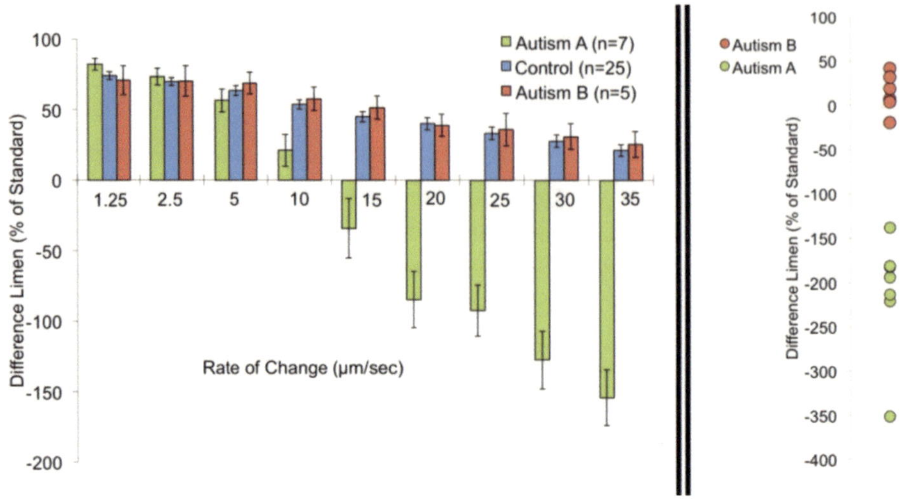

Figure 11. Left Panel: Comparison of data obtained from typically developing controls vs. individuals with autism. Note that at lower rates of stimulus amplitude modulation, all 3 groups behave approximately the same way. As the amplitude modulation rate is increased, the responses of one of the autism groups diverge distinctly from the responses of the other subjects. Note that the negative Weber fraction indicates that the subject responded beyond the matching point of the two stimuli rather than before. Right Panel: Comparison of individual data points from the highest modulation rate displayed in Panel A. Note the clustering of the data points within each of the groups of subjects.

7. Generating an individual CNS profile from multiple measures

A battery of protocols yields multiple parameters that can be used to build a CNS profile of a subject. Since each of the tests are influenced by some mechanisms more than others (e.g., adaptation will influence the evoked cortical response during conditioning prior to a TOJ task, but synchronization of cortical ensembles appears to have the predominant outcome on that task), combining the results from multiple tasks – with each task characterized as an independent vector of performance for some aspect of CNS information processing - would predictably yield a unique individual CNS profile. To fully appreciate the differences between subject populations, we utilized a modern mathematical approach for multi-variable analysis. Quantitative performance of each subject on the battery of *N* sensory tests was treated as localizing a subject in an *N*-dimensional "cortical metrics" space (i.e., an abstract space in which each coordinate axis corresponds to one of the battery's sensory tests). Principal Component Analysis (PCA) was then used to graphically display the test-performance data collected in the different subject populations. Figure 12, for example, was generated using PCA on 8 metrics, and it clearly separates individuals with autism (orange) from TD controls (blue) with a 99% confidence level that the two populations are different (using a t-squared Hotelling test). Our long term goal with this work is to develop metrics that have the requisite sensitivity to reflect the impact that treatments or interventions have. It is anticipated that successful treatments would result in a shift of the autism values towards the more tightly clustered control values.

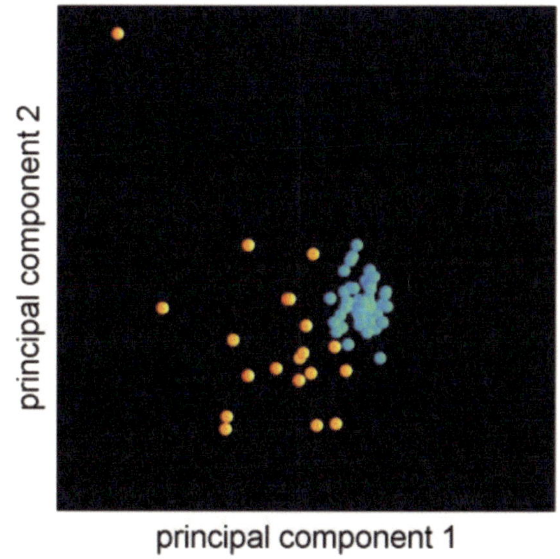

Figure 12. PCA Analysis was used to examine the performance of two populations on 8 differerent metrics. The analysis clearly separates individuals with autism (orange) from TD controls (blue). (99% confidence using a t-squared Hotelling test)

8. Conclusions

Adults with autism exhibit inhibitory deficits that are often manifested in behavioral modifications such as repetitive behaviors and/or sensory hyper-responsiveness. If such behaviors are the result of a generalized deficiency in inhibitory neurotransmission, then it stands to reason that deficits involving localized cortical-cortical interactions – such as in sensory discrimination tasks – could be detected and quantified. This chapter describes recently developed hypothesis driven methods for quantifying metrics of sensory perception based on the neurophysiological principles of cortical modularity. These novel sensory discrimination tests may provide (a) an effective means for biobehavioral assessment of deficits specific to autism and (b) efficient and sensitive measures of change following treatment. The methods could prove to be a useful and efficient way to detect specific neural deficits and monitor the efficacy of pharmacological and/or behavioral treatments in autism.

Author details

Eric Francisco, Oleg Favorov and Mark Tommerdahl

Biomedical Engineering University of North Carolina at Chapel Hill Chapel Hill, NC, USA

References

[1] Casanova, M. F, Buxhoeveden, D. P, Switala, A. E, & Roy, E. Minicolumnar pathology in autism. Neurology (2002). , 58, 428-32.

[2] Mountcastle, V. B. In: Edelman GM, Mountcastle VB, editors. The Mindful Brain. Cambridge, MA: MIT Press; (1978). , 51-100.

[3] Rakic, P. Specification of cerebral cortical areas. Science (1988). , 241, 170-6.

[4] Peters, A, & Yilmaz, E. Neuronal organization in area 17 of cat visual cortex. Cereb Cortex (1993). , 3, 49-68.

[5] Jones, E. G. Varieties and distribution of non-pyramidal cells in the somatic sensory cortex of the squirrel monkey. J Comp Neurol (1975). , 160, 205-67.

[6] Jones, E. G. Anatomy of cerebral cortex: columnar input-output organization. In: Schmitt FO, editor. The Organization of the Cerebral Cortex. Cambridge, MA: MIT Press; (1981). , 199-235.

[7] Lund, J. S. Spiny Stellate Neurons. In: Jones EG, editor. Cerebral Cortex. New York: Plenum Press; (1984). , 255-308.

[8] Favorov, O. V, & Diamond, M. E. Demonstration of discrete place-defined columns--segregates--in the cat SI. J Comp Neurol (1990). , 298, 97-112.

[9] Favorov, O, & Whitsel, B. L. Spatial organization of the peripheral input to area 1 cell columns. I. The detection of'segregates'. Brain Res (1988). , 472, 25-42.

[10] Favorov, O. V, & Kelly, D. G. Minicolumnar Organization within Somatosensory Cortical Segregates: I. Development of Afferent Connections. Cereb Cortex (1994). , 4(4), 408-427.

[11] Tommerdahl, M, Favorov, O, & Whitsel, B. L. Optical imaging of intrinsic signals in somatosensory cortex. Behav Brain Res (2002). , 135, 83-91.

[12] Chiu, J. S, Tommerdahl, M, Whitsel, B. L, & Favorov, O. V. Stimulus-dependent spatial patterns of response in SI cortex. BMC Neurosci (2005).

[13] Chiu, J. Characterization of Minicolumnar Patterns in SI Cortex [Dissertation]. Chapel Hill: University of North Carolina at Chapel Hill; (2006).

[14] Tommerdahl, M, Chiu, J, Whitsel, B. L, & Favorov, O. Minicolumnar patterns in the global cortical response to sensory stimulation. In: Casanova MF, editor. Neocortical Modularity and the Cell Minicolumn. Hauppauge, New York: Nova Science Publishers, (2005).

[15] Tommerdahl, M, Favorov, O. V, & Whitsel, B. L. Dynamic representations of the somatosensory cortex. Neurosci Biobehav Rev. [Review]. (2010). Feb;, 34(2), 160-70.

[16] Casanova, M. F, Buxhoeveden, D, & Gomez, J. Disruption in the inhibitory architecture of the cell minicolumn: implications for autisim. Neuroscientist. (2003). Dec;, 9(6), 496-507.

[17] Tommerdahl, M, Whitsel, B. L, Cox, E. G, Diamond, M. E, & Kelly, D. G. Analysis of the periodicities in somatosensory cortical activity patterns. Society for Neuroscience Abstracts. (1987).

[18] Tommerdahl, M, Favorov, O, Whitsel, B. L, Nakhle, B, & Gonchar, Y. A. Minicolumnar activation patterns in cat and monkey SI cortex. Cereb Cortex. (1993). Sep-Oct;, 3(5), 399-411.

[19] Mccasland, J. S, & Woolsey, T. A. High-resolution 2-deoxyglucose mapping of functional cortical columns in mouse barrel cortex. The Journal of comparative neurology. (1988). Dec 22;, 278(4), 555-69.

[20] Kohn, A, Metz, C, Quibrera, M, Tommerdahl, M. A, & Whitsel, B. L. Functional neocortical microcircuitry demonstrated with intrinsic signal optical imaging in vitro. Neuroscience. (2000). , 95(1), 51-62.

[21] Tommerdahl, M, Tannan, V, Cascio, C. J, Baranek, G. T, & Whitsel, B. L. Vibrotactile adaptation fails to enhance spatial localization in adults with autism. Brain Res. (2007). Jun 18;, 1154, 116-23.

[22] Tannan, V, Whitsel, B. L, & Tommerdahl, M. A. Vibrotactile adaptation enhances spatial localization. Brain Res. (2006). Aug 2;, 1102(1), 109-16.

[23] Simons, S. B, Tannan, V, Chiu, J, Favorov, O. V, Whitsel, B. L, & Tommerdahl, M. Amplitude-dependency of response of SI cortex to flutter stimulation. BMC Neurosci. (2005).

[24] Simons, S. B, Chiu, J, Favorov, O. V, Whitsel, B. L, & Tommerdahl, M. Duration-dependent response of SI to vibrotactile stimulation in squirrel monkey. Journal of neurophysiology. (2007). Mar;, 97(3), 2121-9.

[25] Juliano, S. L, Dusart, I, & Peschanski, M. Somatic activation of thalamic neurons transplanted into lesioned somatosensory thalamus. Brain Res. (1989). Jan 30;, 478(2), 356-60.

[26] Kohn, A, Metz, C, Quibrera, M, Tommerdahl, M. A, & Whitsel, B. L. Functional neocortical microcircuitry demonstrated with intrinsic signal optical imaging in vitro. Neuroscience. (2000). , 95(1), 51-62.

[27] Hussman, J. P. Suppressed GABAergic inhibition as a common factor in suspected etiologies of autism. Journal of autism and developmental disorders. (2001). Apr;, 31(2), 247-8.

[28] Blatt, G. J. GABAergic cerebellar system in autism: a neuropathological and developmental perspective. International review of neurobiology. (2005). , 71, 167-78.

[29] Casanova, M. F, Buxhoeveden, D, & Gomez, J. Disruption in the inhibitory architecture of the cell minicolumn: implications for autisim. Neuroscientist. (2003). Dec;, 9(6), 496-507.

[30] Belmonte, M. K, & Cook, E. H. Jr., Anderson GM, Rubenstein JL, Greenough WT, Beckel-Mitchener A, et al. Autism as a disorder of neural information processing: directions for research and targets for therapy. Mol Psychiatry. [Review]. (2004). Jul;, 9(7), 646-63.

[31] Chao, H. T, Chen, H, Samaco, R. C, Xue, M, Chahrour, M, Yoo, J, et al. Dysfunction in GABA signalling mediates autism-like stereotypies and Rett syndrome phenotypes. Nature (2010). Nov 11;, 468(7321), 263-9.

[32] Brumberg, J. C, Pinto, D. J, & Simons, D. J. Spatial gradients and inhibitory summation in the rat whisker barrel system. Journal of neurophysiology. (1996). Jul;, 76(1), 130-40.

[33] Derdikman, D, Hildesheim, R, Ahissar, E, Arieli, A, & Grinvald, A. Imaging spatio-temporal dynamics of surround inhibition in the barrels somatosensory cortex. J Neurosci. (2003). Apr 15;, 23(8), 3100-5.

[34] Foeller, E, Celikel, T, & Feldman, D. E. Inhibitory sharpening of receptive fields contributes to whisker map plasticity in rat somatosensory cortex. Journal of neuro-physiology. (2005). Dec;, 94(6), 4387-400.

[35] Wirth, C, & Luscher, H. R. Spatiotemporal evolution of excitation and inhibition in the rat barrel cortex investigated with multielectrode arrays. Journal of neurophysiology. (2004). Apr;, 91(4), 1635-47.

[36] Moore, C. I, Nelson, S. B, & Sur, M. Dynamics of neuronal processing in rat somato-sensory cortex. Trends in neurosciences. (1999). Nov;, 22(11), 513-20.

[37] Tommerdahl, M, Whitsel, B. L, & Vierck, C. J. Jr., Favorov O, Juliano S, Cooper B, et al. Effects of spinal dorsal column transection on the response of monkey anterior parietal cortex to repetitive skin stimulation. Cereb Cortex. (1996). Mar-Apr;, 6(2), 131-55.

[38] Tommerdahl, M, Whitsel, B. L, Favorov, O. V, Metz, C. B, & Quinn, O. BL. Responses of contralateral SI and SII in cat to same-site cutaneous flutter versus vibration. Journal of neurophysiology. (1999). Oct;, 82(4), 1982-92.

[39] Grinvald, A. Real-time optical mapping of neuronal activity: from single growth cones to the intact mammalian brain. Annu Rev Neurosci. (1985). , 8, 263-305.

[40] Grinvald, A, Frostig, R. D, Siegel, R. M, & Bartfeld, E. High-resolution optical imaging of functional brain architecture in the awake monkey. Proceedings of the National Academy of Sciences of the United States of America. (1991). Dec 15;, 88(24), 11559-63.

[41] Whitsel, B. L, Kelly, E. F, Quibrera, M, Tommerdahl, M, Li, Y, Favorov, O. V, et al. Time-dependence of SI RA neuron response to cutaneous flutter stimulation. Somatosensory & motor research. (2003). , 20(1), 45-69.

[42] Bredfeldt, C. E, & Ringach, D. L. Dynamics of spatial frequency tuning in macaque J Neurosci. (2002). Mar 1;22(5):1976-84., 1

[43] Celebrini, S, Thorpe, S, Trotter, Y, & Imbert, M. Dynamics of orientation coding in area of the awake primate. Vis Neurosci. (1993). Sep-Oct;10(5):811-25., 1

[44] Das, A, & Gilbert, C. D. Receptive field expansion in adult visual cortex is linked to dynamic changes in strength of cortical connections. Journal of neurophysiology. (1995). Aug;, 74(2), 779-92.

[45] Deangelis, G. C, Anzai, A, Ohzawa, I, & Freeman, R. D. Receptive field structure in the visual cortex: does selective stimulation induce plasticity? Proceedings of the National Academy of Sciences of the United States of America. (1995). Oct 10;, 92(21), 9682-6.

[46] Dinse, H. R, & Kruger, K. Contribution of area 19 to the foreground-background-interaction of the cat: an analysis based on single cell recordings and behavioural experiments. Experimental brain research Experimentelle Hirnforschung. (1990). , 82(1), 107-22.

[47] Pack, C. C, & Born, R. T. Temporal dynamics of a neural solution to the aperture problem in visual area MT of macaque brain. Nature. (2001). Feb 22;, 409(6823), 1040-2.

[48] Pettet, M. W, & Gilbert, C. D. Dynamic changes in receptive-field size in cat primary visual cortex. Proceedings of the National Academy of Sciences of the United States of America. (1992). Sep 1;, 89(17), 8366-70.

[49] Ringach, D. L, Hawken, M. J, & Shapley, R. Dynamics of orientation tuning in macaque primary visual cortex. Nature. (1997). May 15;, 387(6630), 281-4.

[50] Shevelev, I. A, Eysel, U. T, Lazareva, N. A, & Sharaev, G. A. The contribution of intracortical inhibition to dynamics of orientation tuning in cat striate cortex neurons. Neuroscience. (1998). May;, 84(1), 11-23.

[51] Shevelev, I. A, Volgushev, M. A, & Sharaev, G. A. Dynamics of responses of neurons evoked by stimulation of different zones of receptive field. Neuroscience. (1992). Nov; 51(2):445-50., 1

[52] Sugase, Y, Yamane, S, Ueno, S, & Kawano, K. Global and fine information coded by single neurons in the temporal visual cortex. Nature. (1999). Aug 26;, 400(6747), 869-73.

[53] Kohn, A. Visual adaptation: physiology, mechanisms, and functional benefits. Journal of neurophysiology. (2007). May;, 97(5), 3155-64.

[54] Smith, M. A, & Kohn, A. Spatial and temporal scales of neuronal correlation in primary visual cortex. J Neurosci (2008). Nov 26;, 28(48), 12591-603.

[55] Kohn, A, & Whitsel, B. L. Sensory cortical dynamics. Behavioural brain research. (2002). Sep 20;135(1-2):119-26.

[56] Tommerdahl, M, Whitsel, B. L, & Vierck, C. J. Jr., Favorov O, Juliano S, Cooper B, et al. Effects of spinal dorsal column transection on the response of monkey anterior parietal cortex to repetitive skin stimulation. Cereb Cortex. (1996). Mar-Apr;, 6(2), 131-55.

[57] Cannestra, A. F, Pouratian, N, Shomer, M. H, & Toga, A. W. Refractory periods observed by intrinsic signal and fluorescent dye imaging. Journal of neurophysiology. (1998). Sep;, 80(3), 1522-32.

[58] Tannan, V, Simons, S, Dennis, R. G, & Tommerdahl, M. Effects of adaptation on the capacity to differentiate simultaneously delivered dual-site vibrotactile stimuli. Brain Res. (2007). Oct 22.

[59] Tannan, V, Holden, J. K, Zhang, Z, Baranek, G. T, & Tommerdahl, M. A. Perceptual metrics of individuals with autism provide evidence for disinhibition. Autism Res (2008). Aug;, 1(4), 223-30.

[60] Rinehart, N. J, Tonge, B. J, Bradshaw, J. L, Iansek, R, Enticott, P. G, & Mcginley, J. Gait function in high-functioning autism and Asperger's disorder: evidence for basal-ganglia and cerebellar involvement? European child & adolescent psychiatry. (2006). Aug;, 15(5), 256-64.

[61] Takarae, Y, Minshew, N. J, Luna, B, & Sweeney, J. A. Atypical involvement of frontostriatal systems during sensorimotor control in autism. Psychiatry research. (2007). Nov 15;, 156(2), 117-27.

[62] Lacruz, F, Artieda, J, Pastor, M. A, & Obeso, J. A. The anatomical basis of somaesthetic temporal discrimination in humans. Journal of neurology, neurosurgery, and psychiatry. (1991). Dec;, 54(12), 1077-81.

[63] Pastor, M. A, Day, B. L, Macaluso, E, Friston, K. J, & Frackowiak, R. S. The functional neuroanatomy of temporal discrimination. J Neurosci. (2004). Mar 10;, 24(10), 2585-91.

[64] Laasonen, M, Tomma-halme, J, Lahti-nuuttila, P, Service, E, & Virsu, V. Rate of information segregation in developmentally dyslexic children. Brain Lang. (2000). Oct 15;, 75(1), 66-81.

[65] Sanger, T. D, Tarsy, D, & Pascual-leone, A. Abnormalities of spatial and temporal sensory discrimination in writer's cramp. Mov Disord. (2001). Jan;, 16(1), 94-9.

[66] Tinazzi, M, Fiaschi, A, Frasson, E, Fiorio, M, Cortese, F, & Aglioti, S. M. Deficits of temporal discrimination in dystonia are independent from the spatial distance between the loci of tactile stimulation. Mov Disord. (2002). Mar;, 17(2), 333-8.

[67] Tinazzi, M, Frasson, E, Bertolasi, L, Fiaschi, A, & Aglioti, S. Temporal discrimination of somesthetic stimuli is impaired in dystonic patients. Neuroreport. (1999). May 14;, 10(7), 1547-50.

[68] Artieda, J, Pastor, M. A, Lacruz, F, & Obeso, J. A. Temporal discrimination is abnormal in Parkinson's disease. Brain. (1992). Feb;115 Pt , 1, 199-210.

[69] Tommerdahl, M, Tannan, V, Holden, J. K, & Baranek, G. T. Absence of stimulus-driven synchronization effects on sensory perception in autism: Evidence for local underconnectivity? Behav Brain Funct. (2008).

[70] Haist, F, Adamo, M, Westerfield, M, Courchesne, E, & Townsend, J. The functional neuroanatomy of spatial attention in autism spectrum disorder. Developmental neuropsychology. (2005). , 27(3), 425-58.

[71] Herbert, M. R, Ziegler, D. A, Deutsch, C. K, Brien, O, Lange, L. M, & Bakardjiev, N. A, et al. Dissociations of cerebral cortex, subcortical and cerebral white matter volumes in autistic boys. Brain. (2003). May;126(Pt 5):1182-92.

[72] Hollander, E, Anagnostou, E, Chaplin, W, Esposito, K, Haznedar, M. M, Licalzi, E, et al. Striatal volume on magnetic resonance imaging and repetitive behaviors in autism. Biological psychiatry. (2005). Aug 1;, 58(3), 226-32.

[73] Langen, M, Durston, S, Staal, W. G, Palmen, S. J, & Van Engeland, H. Caudate nucleus is enlarged in high-functioning medication-naive subjects with autism. Biological psychiatry. (2007). Aug 1;, 62(3), 262-6.

[74] Rojas, D. C, Peterson, E, Winterrowd, E, Reite, M. L, Rogers, S. J, & Tregellas, J. R. Regional gray matter volumetric changes in autism associated with social and repetitive behavior symptoms. BMC psychiatry. (2006).

[75] Sears, L. L, Vest, C, Mohamed, S, Bailey, J, Ranson, B. J, & Piven, J. An MRI study of the basal ganglia in autism. Progress in neuro-psychopharmacology & biological psychiatry. (1999). May;, 23(4), 613-24.

[76] Voelbel, G. T, Bates, M. E, Buckman, J. F, Pandina, G, & Hendren, R. L. Caudate nucleus volume and cognitive performance: Are they related in childhood psychopathology? Biological psychiatry. (2006). Nov 1;, 60(9), 942-50.

[77] Hardan, A. Y, Girgis, R. R, Adams, J, Gilbert, A. R, Keshavan, M. S, & Minshew, N. J. Abnormal brain size effect on the thalamus in autism. Psychiatry research. (2006). Oct 30;147(2-3):145-51.

[78] Hardan, A. Y, Girgis, R. R, Adams, J, Gilbert, A. R, Melhem, N. M, Keshavan, M. S, et al. Brief Report: Abnormal Association Between the Thalamus and Brain Size in Asperger's Disorder. Journal of autism and developmental disorders. (2008). Feb;, 38(2), 390-4.

[79] Lee, J. E, Bigler, E. D, Alexander, A. L, & Lazar, M. DuBray MB, Chung MK, et al. Diffusion tensor imaging of white matter in the superior temporal gyrus and temporal stem in autism. Neuroscience letters. (2007). Sep 7;, 424(2), 127-32.

[80] Uhlhaas, P. J, & Singer, W. Neural synchrony in brain disorders: relevance for cognitive dysfunctions and pathophysiology. Neuron. (2006). Oct 5;, 52(1), 155-68.

[81] Just, M. A, Cherkassky, V. L, Keller, T. A, Kana, R. K, & Minshew, N. J. Functional and anatomical cortical underconnectivity in autism: evidence from an FMRI study of an

executive function task and corpus callosum morphometry. Cereb Cortex. (2007). Apr;, 17(4), 951-61.

[82] Castelli, F, Frith, C, Happe, F, & Frith, U. Autism, Asperger syndrome and brain mechanisms for the attribution of mental states to animated shapes. Brain. (2002). Aug; 125(Pt 8):1839-49.

[83] Koshino, H, Carpenter, P. A, Minshew, N. J, Cherkassky, V. L, Keller, T. A, & Just, M. A. Functional connectivity in an fMRI working memory task in high-functioning autism. NeuroImage. (2005). Feb 1;, 24(3), 810-21.

[84] Villalobos, M. E, Mizuno, A, Dahl, B. C, Kemmotsu, N, & Muller, R. A. Reduced functional connectivity between and inferior frontal cortex associated with visuomotor performance in autism. NeuroImage. (2005). Apr 15;25(3):916-25., 1

[85] Kana, R. K, Keller, T. A, Cherkassky, V. L, Minshew, N. J, & Just, M. A. Sentence comprehension in autism: thinking in pictures with decreased functional connectivity. Brain. (2006). Sep;129(Pt 9):2484-93.

[86] Cherkassky, V. L, Kana, R. K, Keller, T. A, & Just, M. A. Functional connectivity in a baseline resting-state network in autism. Neuroreport. (2006). Nov 6;, 17(16), 1687-90.

[87] Brown, C, Gruber, T, Boucher, J, Rippon, G, & Brock, J. Gamma abnormalities during perception of illusory figures in autism. Cortex; a journal devoted to the study of the nervous system and behavior. (2005). Jun;, 41(3), 364-76.

[88] Wilson, T. W, Rojas, D. C, Reite, M. L, Teale, P. D, & Rogers, S. J. Children and adolescents with autism exhibit reduced MEG steady-state gamma responses. Biological psychiatry. (2007). Aug 1;, 62(3), 192-7.

[89] Whittington, M. A, Traub, R. D, Kopell, N, Ermentrout, B, & Buhl, E. H. Inhibition-based rhythms: experimental and mathematical observations on network dynamics. Int J Psychophysiol. (2000). Dec 1;, 38(3), 315-36.

[90] Tommerdahl, M, Tannan, V, Zachek, M, Holden, J. K, & Favorov, O. V. Effects of stimulus-driven synchronization on sensory perception. Behav Brain Funct. (2007).

[91] Douglas, R. J, Koch, C, Mahowald, M, Martin, K. A, & Suarez, H. H. Recurrent excitation in neocortical circuits. Science (New York, NY. (1995). Aug 18;, 269(5226), 981-5.

[92] Miller, K. D, Pinto, D. J, & Simons, D. J. Processing in layer 4 of the neocortical circuit: new insights from visual and somatosensory cortex. Curr Opin Neurobiol. (2001). Aug;, 11(4), 488-97.

[93] Bruno, R. M, & Simons, D. J. Feedforward mechanisms of excitatory and inhibitory cortical receptive fields. J Neurosci. (2002). Dec 15;, 22(24), 10966-75.

[94] Alonso, J. M, & Swadlow, H. A. Thalamocortical specificity and the synthesis of sensory cortical receptive fields. Journal of neurophysiology. (2005). Jul;, 94(1), 26-32.

http://dx.doi.org/10.5772/54801

[95] Sun, Q. Q, Huguenard, J. R, & Prince, D. A. Barrel cortex microcircuits: thalamocortical feedforward inhibition in spiny stellate cells is mediated by a small number of fast-spiking interneurons. J Neurosci. (2006). Jan 25;, 26(4), 1219-30.

[96] Cruikshank, S. J, Lewis, T. J, & Connors, B. W. Synaptic basis for intense thalamocortical activation of feedforward inhibitory cells in neocortex. Nature neuroscience. (2007). Apr;, 10(4), 462-8.

[97] Zhang, Z, Zolnoun, D. A, Francisco, E. M, Holden, J. K, Dennis, R. G, & Tommerdahl, M. Altered central sensitization in subgroups of women with vulvodynia. Clin J Pain (2011). Nov-Dec;, 27(9), 755-63.

[98] Zhang, Z, Francisco, E, Holden, J. K, Dennis, R. G, & Tommerdahl, M. Sensory information processing in the Aging population. Frontiers Aging Neurosci. (2011).

[99] Francisco, E, Holden, J, Zhang, Z, Favorov, O, & Tommerdahl, M. Rate dependency of vibrotactile stimulus modulation. Brain Res. (2011). Sep 30;, 1415, 76-83.

[100] Tamas, G, Lorincz, A, Simon, A, & Szabadics, J. Identified sources and targets of slow inhibition in the neocortex. Science (New York, NY. (2003). Mar 21;, 299(5614), 1902-5.

Chapter 4

Electrophysiology of Autism

Kristina L. McFadden and Donald C. Rojas

Additional information is available at the end of the chapter

http://dx.doi.org/10.5772/54770

1. Introduction

The purpose of this is to give an overview and summary of findings for event-related potentials, event-related spectral perturbances and spontaneous electrical activity findings in ASD to date. The topic of epilepsy, although relevant to the electrophysiology of autism, is beyond the scope of this chapter. Interested readers are directed to other, recent reviews on this topic [1-4].

1.1. EEG and MEG in Autism Spectrum Disorders (ASD)

Electroencephalography (EEG) and magnetoencephalography (MEG) are noninvasive methods of measuring activity in the brain. Both methods have been used extensively in autism spectrum disorders (ASD), revealing differences in cognitive processing in individuals with ASD compared to typically developing controls. Both EEG and MEG measure synchronized neural activity in the brain with excellent temporal resolution, with EEG measuring electrical potentials on the scalp and MEG measuring the magnetic field outside the head produced by current flow in the brain [5]. Processes in ASD that are commonly investigated with EEG and MEG include auditory processing, face processing, visual processing, speech and language, and social cognition. EEG and MEG can be used to assess simple sensory processing as well as higher-order cognitive processing and can often reveal information that is unavailable from behavioral tasks. Many studies have found differences in brain activity measured by EEG and/or MEG in spite of equivalent performance between ASD and control groups. As such, while behavioral performance may appear normal, measures of brain activity may reveal compensatory processes masking abnormal brain activity [6].

A commonly used measure is the event-related potential (called the event-related field in MEG), which is a waveform created by averaging brain activity recorded with EEG or MEG across multiple trials of a particular task. As such, ERPs are time-locked to the presented stimuli. ERPs have made significant contributions to the current understanding of ASD and have been used extensively as measures of brain function in ASD and in understanding behavioral components of the disorder. In addition to ERPs, the frequency content of brain activity measured by EEG and MEG can be investigated by focusing on event-related spectral perturbances (ERSPs). These measures have become increasingly common as more sophisticated analytical methods are developed and allow for identification of spectral changes that are both phase-locked and non-phase-locked to stimuli.

2. Event Related Potentials (ERP) and Event-Related Spectral Perturbances (ERSP)

Event-related potentials (ERPs), called the event-related field (ERF) in MEG, are created by averaging the response to stimuli across multiple trials. The type of stimuli and experimental paradigm used determine the nature of the observed ERP components and the use of common ERP naming conventions allows for results to be compared across studies. ERPs serve as physiological markers thought to reflect processing of both sensory and higher-level cognitive information and are defined based on polarity (positive or negative) and latency (timing post-stimulus). For example, the N100 is a negative potential (in EEG; polarity is not reflected in MEG) that is seen around 100 ms post-stimulus (see Figure 1). Components are named based on the time at which they occur (e.g., 100 ms) or the order in which they are seen. As such, the N100 may also be referred to as the N1, as it is the first negative component seen in the waveform. To indicate that it is a magnetic event-related field (ERF), the N100 is called the M100 when measured with MEG. Latency of an ERP/ERF indicates the time course of the neural activity, while amplitude is thought to reflect the allocation of neural resources necessary for a particular task [7, 8].

ERP waveforms from a commonly used paradigm, the auditory oddball task, are shown in Figure 1. In this task, study participants listen to a series of sounds in which there are both infrequent ("rare") and frequent ("standard") stimuli. In the example in Figure 1, rare stimuli were 1000 Hz tones (25% of stimuli) presented among the standard stimuli consisting of 500 Hz tones (75% of stimuli). This task elicits a clear N100 response to both types of stimuli, but the P300 component is much larger to rare stimuli compared to standard stimuli. This type of paradigm also allows for the detection of the mismatch negativity (MMN) component, which is a response elicited by changes in stimuli. This component is derived by subtracting the standard waveform from the rare waveform, as can be seen in Figure 1.

Over the past 10 years, while signal averaged ERP analyses have continued to develop with increased sophistication, there has also been increased awareness of the loss of information in MEG and EEG due to time-domain averaging. Partly motivated by time-frequency transformations and by the development of metrics to assess the underlying assumption of time/phase-locking of the response to a stimulus, researchers have focused more recently on stimulus-related change in the frequency content of EEG/MEG data. With time-domain averaging, any jitter in the phase of the EEG response with respect to the onset of the stimulus or response causes a reduction in the amplitude of the ERP, or spreading of the amplitude, and with enough jitter, will reduce the ERP to the noise floor. ERP phenomena are therefore reliant on a relatively high degree of phase-locking (often called time-locking in the ERP literature) and high inter-trial phase consistency. Responses that are highly phase-locked are considered evoked responses. There are, however, legitimate stimulus-related EEG signals that appear to be inherently non-phase-locked (i.e., the signal change is present, but varies from trial to trial). Such non-phase-locked events are known as induced responses. While averaging such responses in the time domain tends to eliminate them, averaging of induced responses can be achieved by first transforming the single trial data into the frequency domain, then averaging. Figure 2 illustrates two types of ERSP, one highly phase-locked and one not, from MEG experiments.

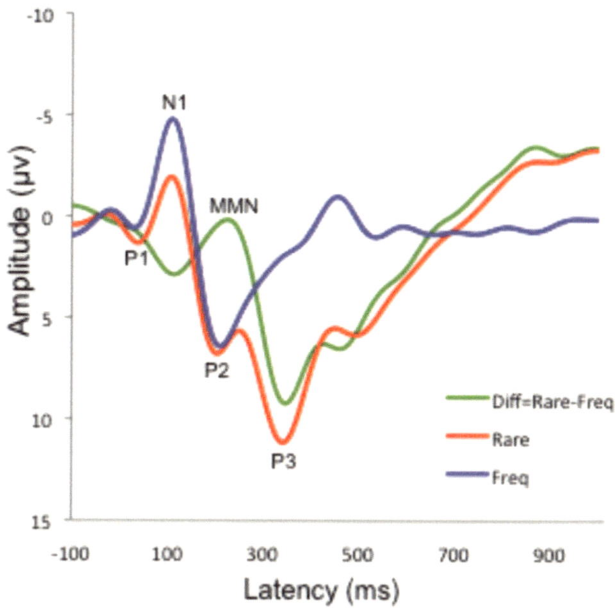

Figure 1. Event-related potential (ERP) waveforms. Responses to an auditory oddball task, in which infrequent (rare) stimuli consisting of 1000 Hz tones (70 ms duration; 25% of stimuli) were presented among frequent (freq) 500 Hz tones (70 ms duration; 75% of stimuli) binaurally. Participants responded to rare tones with a button press. The ERP response to the rare tones is shown in red, with the response to the frequent tones in blue. Common ERP components (P1, N1, P2, P3) are labeled. The mismatch negativity (MMN) is shown in green and is derived by subtracting the response to the standard tones from the response to the rare tones. Data shown are adapted from McFadden et al. [177].

Various analyses of spectral changes evoked or induced by stimuli or associated with responses have been collectively termed event-related spectral perturbances (ERSP) by Makeig and colleagues [9]. The term ERSP encompasses the earlier and still often used terms event-related desynchronization (ERD) and event-related synchronization (ERS), terms that describe stimulus-related decreases and increases in spectral power, respectively. ERSP, ERD and ERS phenomena can be evoked (phase-locked), induced (non-phase-locked), or both.

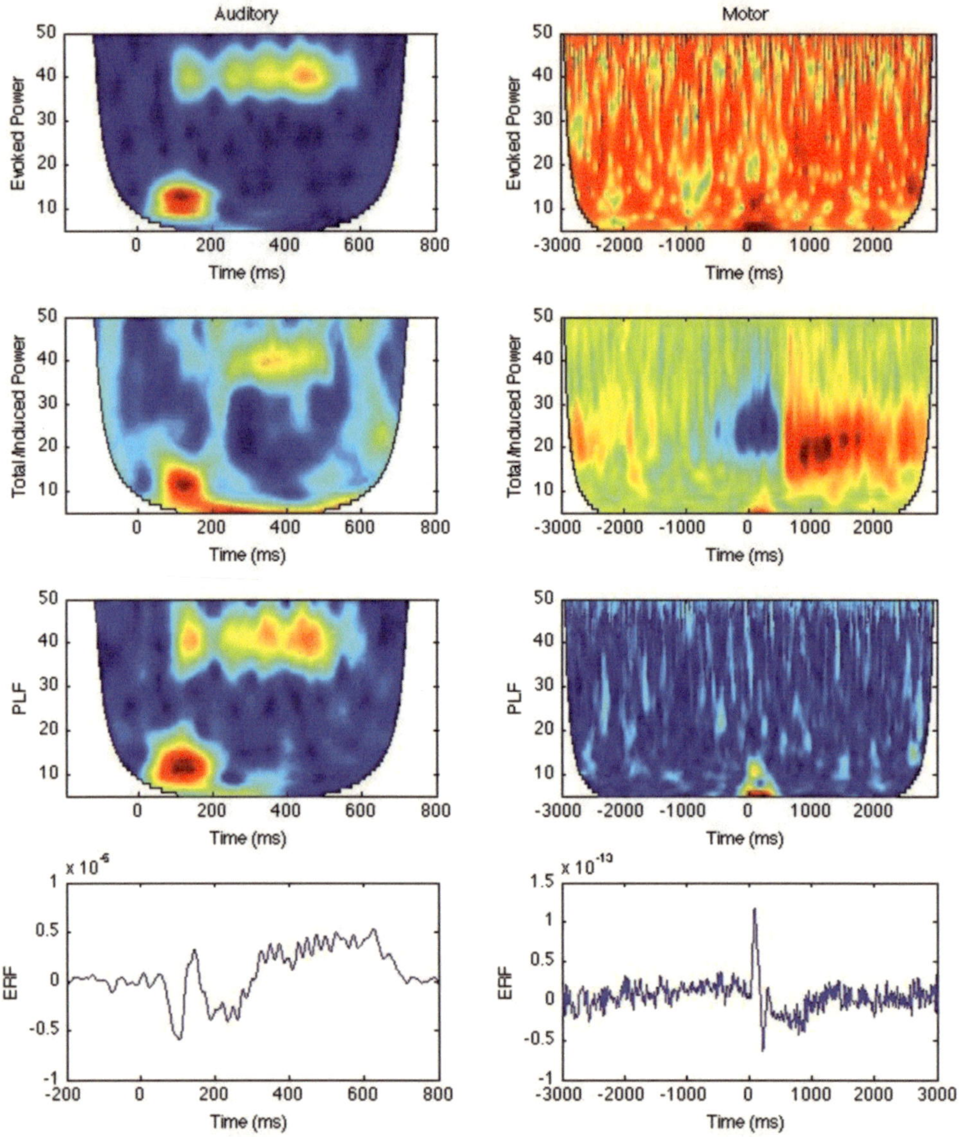

Figure 2. Event-related spectral perturbances (ERSP). Auditory response to 500 ms, 40 Hz modulated stimulus (Auditory, left column) and sensorimotor response to unilateral index finger movement (Motor, right column) are shown. Top row = evoked power, relative to prestimulus baseline. Second row = total, or induced power, relative to baseline. Third row = phase-locking factor (PLF). Bottom row = time-domain averaged evoked field (ERF). For rows 2 and 3, ERS is shown in red colors and ERD in blue colors. For the auditory stimulus, the baseline period for relative power and DC offset is the -200 to 0 ms period preceding the stimulus. For the motor ERSP, the baseline period is -3000 to -2000 ms. For the auditory steady-state response, which is highly phase-locked (see third row), the gamma response at 40 Hz is clearly evident in the evoked power, but can also be seen in the time-domain average. The motor beta ERD and ERS responses, which are not phase-locked, are best visualized in the induced, or total power. Note that the low frequency motor evoked field, clearly observed at time 0 in the motor response, is phase-locked and seen in the PLF, evoked and total power.

http://dx.doi.org/10.5772/54770

3. Event-Related Potentials in ASD

Several ERP components have been extensively investigated in ASD and the following sections will detail the cognitive processes thought to be involved in each, as well as how these components have been used to assess cognition in ASD.

3.1. N100/M100

The N100 and its magnetic equivalent, the M100, is usually seen between 60 and 160 ms [10] and is thought to reflect processing of basic sensory (mainly auditory) features [11-14]. The N100/M100 is thought to be primarily generated by auditory cortex [15-17]. While the N100 is often broken into multiple components, the M100 (measured by MEG) is a more straightforward measure of activity from primary auditory cortex [11, 18, 19]. The N/M100 is often elicited by auditory stimuli and a common paradigm used to measure the N/M100 is the auditory oddball paradigm, in which participants hear infrequent ("rare") sounds among frequent ("standard") sounds (see Figure 1). However, the N/M100 can also be seen in response to stimuli in other sensory domains, such as visual or somatosensory stimuli [11].

Many studies have investigated differences in N/M100 latency in ASD, although results are somewhat inconsistent. While most studies have found no differences in N/M100 amplitude between individuals with ASD and typically developing controls [20-28], one study found individuals with ASD to demonstrate decreased amplitude compared to controls [29] and two others found increased amplitude in ASD [30, 31]. Strandburg et al. [30] suggested the reason for their finding of increased N100 amplitude in ASD could be due to the complexity of the stimuli used. Their study employed a number of information processing tasks that were more difficult than the simple auditory oddball paradigms used in many of the studies finding either decreased or equivalent N100 amplitude in ASD groups. Similarly, Oades et al. [31], who also found increased N100 amplitude in individuals with ASD compared to controls, also hypothesized that the tasks used may have been too difficult for the ASD population included in the study. While they used an auditory oddball task, participants were required to attend and respond to stimuli throughout the task, while many others employ a passive task. Oades et al. also found the ASD group in their study to show worse performance during the task, indicating that the increased N100 amplitude may reflect difficulties in attending to stimuli. Attention deficits are often seen in ASD [32-34] and the N/M100 is thought be impacted by early attentional processes [7, 11, 35, 36].

Interesting results have also been found in regard to N/M100 latency. While some studies identified no differences in N/M100 latency between individuals with ASD and typically developing controls [21, 26, 27, 37, 38], others have found either delayed [14, 28, 39-42] or earlier N/M100 latency in ASD [20, 25, 31]. Given the consistency of the finding, delayed N/M100 latency has been suggested as a potential biomarker for ASD [14, 42]. It is thought that this could reflect abnormal auditory processing in individuals with ASD, at the level of simple sensory encoding. Consistent with this hypothesis, many studies have found evidence of abnormalities in auditory processing in ASD [6, 12, 43, 44]. As with abnormal N/M100 amplitude, this abnormal latency could reflect impairments in early attentional processes in

ASD. Additionally, it has been suggested that N/M100 latency delays in ASD may be a marker of language dysfunction in this population [14, 42].

The differences among studies are not surprising, given the myriad of differences among study designs. While some studies focused on low-functioning individuals with ASD, other focused on a high-functioning population. Additionally, disparities between studies can be attributed to differences in age groups studied and differences in experimental paradigms used. While the N/M100 is reliable in adults, it has not been found to be as consistent in children, due to developmental aspects of the response. The N/M100 response is sometimes not evident in young children [45] and has been found to show increased amplitude and decreased latency with increasing age [17, 18, 38, 46], thought to be partly related to ongoing myelination of white matter [18, 47]. This reduced latency in aging may also be delayed in children with ASD [40].

Additionally, Schmidt et al. [48] found that children with ASD fail to show the anticipated asymmetry in M100 generator location. Typically developing controls show M100 generator location asymmetry such that the location in the brain is more anterior in the right hemisphere than the left hemisphere [49]. However, Schmidt et al. found children with ASD to demonstrate an absence of this asymmetry [48]. Schmidt et al. also found the degree of asymmetry to be associated with behavioral measures of language function, suggesting a relationship between the M100 and language abilities. This is noteworthy as language impairments are a core symptom of ASD [50]. As such, it has been suggested that M100 abnormalities may be related to language impairments rather than specifically to ASD [48]. However, Roberts et al. found that children with ASD demonstrated M100 latency delays regardless of whether they had language impairments or not [14], suggesting this response to be specifically related to ASD rather than simply to language impairment. Supporting this, Roberts et al. have also found that children with specific language impairment who do not have ASD do not show M100 latency delays [51].

3.2. Mismatch negativity (MMN/MMF)

The MMN (or the mismatch field (MMF), its MEG equivalent) is an early component typically seen between 150-250 ms post-stimulus and is elicited by changes in stimuli [7, 8, 13, 52, 53]. The magnitude of perceptual difference between stimuli predicts the amplitude of the MMN response [13, 52]. While the MMN is modulated by attention, it can be elicited by stimuli differences even in the absence of attention, indicating that it is an early marker of pre-attentive detection of change [7, 8, 13, 53]. The MMN is elicited by presenting occasional ("rare") stimuli interspersed among frequent ("standard") stimuli, as in an auditory oddball paradigm [7, 54]. The MMN is commonly measured by creating a difference wave, using the "rare" minus "standard" waveforms (see Figure 1) [12, 24, 52, 53]. While the MMN is most frequently studied in response to auditory stimuli, it has also been elicited in other sensory modalities [53, 55]. The MMN has been implicated as indicating a number of brain functions, such as auditory discrimination [6, 52, 53, 55, 56], duration of sensory memory [52, 53, 55], abnormal auditory perception [55], involuntary attention switching [6, 53, 55], cognitive function [55], illness progression [55, 57], and treatment outcome [45, 55].

http://dx.doi.org/10.5772/54770

Given that individuals with ASD are thought to have abnormalities in auditory processing [6, 12, 43, 44], the MMN can be a useful tool in this population as it can be used to assess various components of auditory processing. The MMN can be used as a measure of the ability to discriminate between various simple or complex sounds [52], which could identify areas of processing potentially leading to the speech and language difficulties common in ASD. In addition to being elicited by changes in simple stimuli, such as tones, the MMN can also be seen in response to higher-order discrimination paradigms, such as detecting grammatical errors and discriminating between words and pseudo-words [53]. Since the MMN can be elicited in the absence of attention, this also makes it a particularly useful measure for ASD populations [52]. Attention deficits are often seen in ASD [32-34] and if attention is necessary for measurement of a particular component, this can present challenges. The MMN provides a way of avoiding those challenges, as it can be measured while participants are reading a book or watching a movie. It has also been suggested that the MMN be used for early identification of hearing disorders [52] as it can be seen in populations as young as newborn infants [58]. It stands to reason that this component could also be used to assess auditory processing deficits in children with ASD who have not yet acquired language capabilities.

As with other components, MMN/MMF findings in ASD have been somewhat inconsistent. While some have found amplitude to be increased in individuals with ASD compared to typically developing controls [20, 59], others have found amplitude reductions [6, 26, 59, 60] or amplitude measures comparable to controls [24, 34]. In part, these differences depend on the stimulus used; changes in pitch have shown both increased [20, 59] and decreased amplitude [6, 26], and changes in duration have demonstrated reduced MMF amplitude in ASD [59]. Indeed, Lepisto et al. [59] found increased MMN amplitude in children with ASD to pitch changes, but decreased MMN amplitude for changes in duration of stimuli. MMN latency has similarly be found to be both shorter [45], longer [60-62], or comparable between individuals with ASD and controls [24]. Again, this could be due to the stimuli used (e.g., tone stimuli vs. speech stimuli, discrimination between different pitches vs. different durations). Another potential reason for disparities between studies is differences in the ASD populations included. While some studies have focused on MMN/MMF in low-functioning individuals with ASD [20, 26], others have focused on high-functioning individuals [24, 34].

These MMN/MMF differences seen in ASD could indicate impairments in the ability to discriminate between stimuli [6, 53] or impairments in sensory memory [53, 55]. Additionally, since some studies have found attention attenuates group differences in MMN [6, 24], abnormalities in MMN amplitude and/or latency could result from attentional deficits. Dunn et al. [6] addressed the impact of attention on these differences in children with ASD compared to control children. They found that when children were not attending to stimuli, children with ASD demonstrated reductions in MMN amplitude compared to control children. However, when children were instructed to actively attend to stimuli, MMN amplitude was similar between groups. This is consistent with results from Kemner et al. [24], finding similar MMN amplitude in children with ASD compared to controls during a task in which they were actively attending to stimuli. While attention is not necessary to elicit the MMN, attention does modulate the component [63, 64]. As the MMN reflects the ability to automatically switch attention, this could be one aspect of attention that is impaired in ASD; it could be that individuals with ASD need to specifically attend to stimuli to discriminate unanticipated changes, while typically developing individuals are able to detect these changes automatically [6].

It has been suggested that these impairments in auditory discrimination, auditory sensory memory, and attention switching may contribute to the development of language impairments in ASD [6, 43, 47, 59]. While Dunn et al. [6] did not find MMN amplitude to be related to language abilities in children with and without ASD, Roberts [47, 62] has suggested that the MMF latency may serve as a biomarker of language impairment in ASD. Alternatively, it has been suggested that rather than being deficient in auditory discrimination, individuals with ASD may have enhanced auditory discrimination skills for particular stimulus features such as pitch [55, 56, 59, 65]. This is consistent with studies finding increased MMN amplitude [20, 59] or shorter latency [45] to pitch discrimination tasks. It has been suggested that enhanced auditory discrimination skills could also lead to language impairments, through difficulties focusing on relevant details and ignoring irrelevant details when listening to speech sounds [56]. This enhancement has also been hypothesized to potentially underlie auditory hypersensitivity seen in ASD [56, 59]. Indeed, delays in MMF latency seen in children with ASD compared to typically developing controls were found to be greater in children with language impairments [62].

3.3. N170/M170

The N/M170 component is thought to be important in face-specific processing [66-70] and appears to be a sensitive measure of impairment in the early stages of face processing [67, 71, 72]. This component is seen around 170 ms post-stimulus [66, 71, 73]. While it is also elicited to non-face stimuli [72], the N170 shows the shortest latency and greatest amplitude in response to faces [71-75]. The N170 is still seen when faces are inverted, albeit with delayed latency and greater amplitude [71, 73, 74], suggesting that the N170 is involved in simply recognizing something as a face rather than attempting to recognize a particular face [66, 71, 73]. Supporting this, viewing familiar or repeated faces does not result in shorter N170 latency [67, 71, 76, 77]. Of note, the N170 responds to viewing human eyes with the greatest amplitude and shortest latency, leading some to hypothesize that this component is specifically related to eye detection [66, 76]. However, others suggest that the N170 is not specific to faces at all, but reflects expert processing and is seen in response to human faces given that most people are experts in the recognition of faces [67, 78-80].

Studies have found abnormal amplitude and latency of the N170 component in individuals with ASD, but results have been inconsistent [67, 81]. McPartland et al. [72] found adolescents and adults with ASD to show delayed N170 latency in response to face stimuli compared to a control group, but this group difference was not seen in response to objects. While other studies have also seen this N170 latency delay in adults with ASD compared to controls [82-84], others have failed to find a group difference [68, 69]. In some studies, this reduction in N170 latency in ASD was found to generalize across face and non-face stimuli, suggesting slower general processing speed rather than face-specific processing deficits [80]. In addition, studies have found control adults to demonstrate the typical latency delay to inverted faces compared to upright faces, but find this effect to be attenuated in ASD [72, 81]. This lack of an inversion effect has been suggested to reflect less sensitivity to the configural properties of faces [72] and an atypical approach to face processing in ASD [85]. That individuals with ASD do not show the typical additional processing in response to inverted faces supports the theory that individuals with ASD may process the component parts of faces individually [67, 85]. Rather than necessarily reflecting a deficit in global processing, it has been suggested that this reflects individuals with ASD demonstrating superior processing of facial features [67], which has been previously demonstrated [86].

Reduced N170 amplitude in adults with ASD compared to controls has also been seen in some studies [69, 83, 87], but not others [72, 84, 88]. It has been suggested that group differences in N170 amplitude are only seen when task demands are high [81], perhaps reflecting an impact of attention on the component, as attention is often impaired in ASD [32-34]. Churches et al. [87] found N170 amplitude to increase in control adults when asked to actively attend to faces, but failed to see a similar attention-related increase in amplitude in adults with ASD. They hypothesize that face processing is enhanced by attention in typically developing individuals, but that individuals with ASD may not actively attend to faces in the same way and thus do not benefit from attention-enhanced face processing. However, this theory requires further investigation as other studies have found that directing attention to faces appears to result in N170 latency being "normalized" in ASD, with no group differences seen between ASD and controls when attention was directed to the eye region of the face stimuli [68]. Churches et al. [81] also found that the difference in N170 amplitude between individuals with ASD and controls was larger when stimuli were non-face-like objects compared to the group difference seen for faces or face-like objects. This could suggest that the face detection system in those with ASD is perhaps more specific to faces and less capable of generalization to non-face objects. Churches et al. [81] suggest that this reduced generalization may contribute to social impairments as individuals with ASD may be less able to generalize face processing across different circumstances (e.g., different lighting, distances, or orientations).

It has long been thought that individuals with ASD have deficits in face processing [67, 75], but it is as of yet unclear if this deficit is during early processing or at a later cognitive stage of face processing [80]. Face processing impairments in ASD include abnormal eye contact [75, 89], reduced attention to faces [75], deficits in face recognition [75, 76, 81], and impairments in social orienting [76]. These impairments are thought to be early indicators of ASD, since many can be seen by age 2 to 3 [75, 76, 85]. It has also been suggested that reduced social motivation in individuals with ASD may lead to reductions in eye contact and social orienting [67, 76]. Conversely, reduced eye contact early in development is thought to impact development of social communication skills in ASD [75, 90]. Eye contact is a crucial component of social communication and a lack of eye contact is included in standard diagnostic criteria for autism spectrum disorders [50]. Atypical patterns of eye contact in ASD can be seen as early as the first year of life [90-93]. However, it has recently been suggested that face processing deficits in ASD may be overestimated in the literature [67, 80] and that rather than face-specific processing being impaired in ASD, difficulties in processing faces may be related to broader deficits in visual processing [70, 94].

Differences in results across studies could stem from a multitude of reasons, such as the range of function of participants in differing studies (e.g., varying IQ and verbal abilities), the type of control groups used in different studies (e.g., matched by age, matched by verbal ability), and the variability between tasks used in each study (e.g., the type of objects compared to faces, task difficulty) [68, 70]. Age can also influence measures of the N170, since this component is not fully developed until adolescence or adulthood [70, 83, 95].

3.4. P300/M300

The P/M300 is a later component observed around 300 ms post-stimulus onset and is thought to reflect higher order cognitive processing compared to the earlier components. The P300 is often broken into two subcomponents reflecting different cognitive processes, the P3a and P3b subcomponents [96, 97]. The P3a is thought to indicate orienting to environmental changes and response to novelty that could underlie attentional switching [13, 98], and is elicited by rare and task-irrelevant stimuli [13]. The P3b is seen when a participant is responding to a task-relevant target stimulus [12, 13] and as such, is thought to reflect task-related activity and memory updating [98-100]. When the component is referred to as simply the P3 or P300, this is usually the P3b element of the component [7] (as such, this chapter will refer to the P3b component as the P300). A task commonly used to elicit these components is the "oddball" task, in which infrequent ("rare") stimuli are presented among frequent ("standard") stimuli (see Figure 1). The P300 can be elicited by visual, auditory, or somatosensory stimuli [12].

A number of studies have found individuals with ASD to show reduced P300 amplitude compared to controls in response to a variety of auditory stimuli such as clicks, tones, and phonemes [12, 21, 27, 29, 31, 37, 101-103]. This finding is less consistent for visual stimuli, with some studies finding diminished P300 amplitude [103, 104] and others not finding this amplitude reduction [29, 39, 102, 105]. Reductions in amplitude have been suggested to represent individuals with ASD having difficulties attaching significance to the target stimulus [31]. While most studies have found either reduced or equivalent P300 in individuals with ASD, Strandburg et al. [30] found adults with ASD to show increased P300 amplitude to both a continuous performance task (requiring a button press when the same visually-presented number appears on two consecutive trials) and an idiom recognition task (requiring participants to distinguish meaningful idiomatic phrases from nonsensical phrases). However, interpretation of this finding is difficult because while the ASD group performed as well as the control group on the continuous performance task, they performed worse than controls on the idiom recognition task. Given this difference in behavioral performance across the two tasks, and that the two tasks target very different skills, it may be that the amplitude increase in the two different tasks reflects separate cognitive processes. However, given that these tasks are more difficult than most commonly used oddball tasks, the group effect seen here may be a result of the challenging nature of the tasks. As with other components, differences between studies could be due to multiple factors, such as intellectual abilities of participants, matching of control participants to ASD participants, age range included, sensory modality of the stimuli, and the task used to elicit the response. Most studies have not found group differences in P300 latency [21, 22, 37, 101-103], although latency delays in individuals with ASD have been seen in response to novel distractor stimuli in visual target-detection tasks [33, 39].

Interestingly, Salmond et al. [106] found P300 amplitude to be reduced in low-functioning individuals with ASD, but not in high-functioning individuals with ASD. This suggests a relationship between P300 abnormalities and cognitive function. As mentioned, this could also explain discrepancies among other studies, as levels of cognitive abilities between studies are variable. However, reductions in P300 amplitude in ASD have also been seen despite normal behavioral task performance [21, 29, 37], suggesting that individuals with ASD may use compensatory cognitive processing strategies to achieve normal behavioral performance. Others have suggested that abnormalities in the P300 component seen in ASD are due to attentional difficulties, commonly seen in ASD [32-34]. Whitehouse & Bishop [107] found children with ASD to demonstrate reduced P300 amplitude compared to controls when not attending to stimuli, but found that this reduction was attenuated when children were asked to attend to stimuli. This suggests that attention may help to normalize the P300 response in children, suggesting that the deficit may be in automatic processing. It has also been suggested that P300 abnormalities in ASD may be modality-specific, due to abnormalities being seen most consistently to auditory stimuli [30]. It has been suggested that deficits in auditory processing in ASD contribute to impairments in such things as language impairment [69, 98]. Since these P300 abnormalities have also been seen in response to visual stimuli, albeit inconsistently, it has also been hypothesized that visual processing deficits may lead to other social impairments in ASD, such as reading emotions and face processing [98].

3.5. N400/M400

The N/M400 component peak is evident around 400 ms post-stimulus and has been shown to reflect semantic aspects of word and sentence processing [108-111]. As with the P/M300, this component reflects higher order cognitive processing compared to earlier components (e.g., N/M100, MMN/MMF). N/M400 amplitude increases when a stimulus deviates from context, as when a word is not predicted by the surrounding context [13, 111-113]. The larger N/M400 amplitude in response to unexpected stimuli compared to expected stimuli is referred to as an N/M400 effect, or a priming effect. Priming refers to a change in the response to a stimulus as a result of a related preceding stimulus, which is thought to reflect automatic processes involved in such things as memory, category recognition, and language [114, 115]. In semantic priming, responses to words preceded by semantically related words will be enhanced compared to those preceded by unrelated words or nonsense words [112, 114]. Similarly, if a word is incongruent with the preceding sentence (e.g., the boy climbed to the top of the "banana"), it becomes more difficult to integrate the word into the context of the sentence, leading to an increase in N/M400 amplitude [110, 116, 117]. As such, the N/M400 can be used to measure things such as semantic integration and degree of expectancy, which can be used as an indication of language processing [28, 116].

Given that it is a measure of semantic processing, the N/M400 response is of interest in ASD because language impairment is a core symptom of ASD [50]. Many studies have found attenuated semantic priming effects in individuals with ASD compared to controls [28, 113, 118-120].

Dunn et al. [28, 118] showed a classic N400 effect in typically developing control children, in which the N400 was increased in response to words not matching the given category (e.g., animals) compared to the N400 response to words matching the stated category. Children with ASD did not demonstrate this N400 effect; that is, they did not show a difference in N400 amplitude between words that were congruent and incongruent with the category [28, 118]. Fishman et al. [113] found similar results using a more typical N400 paradigm, in which responses to sentences ending with congruent/expected words (e.g., kids learn to read and write in "school") were compared with those to sentences ending in incongruent/unexpected words (e.g., kids learn to read and write in "finger"). In this study, individuals with ASD also demonstrated a lack of an N400 effect; that is, no difference in N400 responses was seen between congruent and incongruent words, although this effect was seen in typically developing controls [113]. Pijnacker et al. [117] also found a lack of the N400 effect in adults with ASD in a sentence context task. McCleery et al. similarly found the N400 effect to be missing in children with ASD during a verbal task (auditory words that matched or did not match a presented picture), but found the N400 effect to be no different from that seen in typically developing children during an environmental sound paradigm involving non-linguistic stimuli (environmental sounds that matched or did not match a picture, e.g., picture of a car and either the sound of a car engine starting (match) or a ball bouncing (mismatch) presented with the picture). This suggests that the deficit may be specific to verbal rather than non-verbal stimuli.

This absence of a priming effect in ASD is suggested to reflect deficits in semantic processing [28, 113, 118, 119]. These deficits could be at multiple points in processing, including attention to stimuli, discriminating between stimuli, and integration of context [13, 110, 116, 117]. It has been suggested that the lack of the N400 effect may indicate that individuals with ASD may fail to use context, or may use it inefficiently [28, 117]. This is supported by other studies suggesting that individuals with ASD have difficulties using contextual information in language tasks [121, 122]. Alternately, it could indicate a lack of semantic representation in individuals with ASD or reduced activation and/or connectivity of brain circuits responsible for semantic processing [28, 113, 119].

It has also been suggested that the reduced or absent N/M400 seen in ASD may be reflecting general verbal or cognitive abilities. Indeed, some studies have also seen impaired behavioral performance in individuals with ASD in addition to a lack of the N400 effect [28, 113]. However, in these studies, ASD groups were not matched to controls on cognitive measures and/or showed reduced cognitive abilities compared to controls. Pijnacker et al. [117] matched control participants to ASD participants on intelligence scores and found individuals with ASD to perform comparably to controls behaviorally, but still found them to lack the N400 effect. Similarly, other studies have also found intact behavioral performance in individuals with ASD despite a reduced or missing N400 effect [118, 119]. This suggests that the lack of an N400 effect is not simply due to a lack of semantic comprehension or task understanding. Rather, the abnormal N400 effect may reflect a lack of automatic semantic processing in individuals with ASD [117, 119]. As such, individuals with ASD seem to be able to effectively use context to arrive at a normal behavioral response, but may require additional effort or more purposeful processing compared to typically developing controls [117]. Findings from McCleery et al.

http://dx.doi.org/10.5772/54770

[119] suggest that this may be specific to certain types of processing. That the reduced N400 effect was not seen in response to nonverbal relationships (i.e., relationships between environmental sounds and pictures) indicates that this may reflect a less effortful form of processing for individuals with ASD. Therefore, while individuals with ASD may have intact automatic processing for some types of stimuli (e.g., nonverbal), automatic processing of semantic information may be specifically impaired. The need for more effortful processing of semantic information that may be more automatic for typically developing children could contribute to the development of additional language and social impairments [119].

While many studies have found the N400 effect to be reduced or absent in ASD [28, 113, 118, 119], other studies have demonstrated an intact effect. Braeutigam et al. [123] found an M400 effect in adults with ASD in response to incongruent compared to congruent sentence endings, using MEG. Although both groups (adults with ASD and a group of control adults) showed an increased M400 response to incongruent compared to congruent sentence endings, there were differences in lateralization. While the control group showed a bilaterally increased response to incongruent endings, the ASD group showed a stronger right hemisphere response. Braeutgiam et al. suggest that this could be due to those in the ASD group having weaker expectancy regarding the sentence endings compared to the control group, as the left hemisphere is thought to be biased towards an expected meaning [123].

4. Resting-state electrophysiology in ASD

One of the most prominent features of resting EEG/MEG in autism is the presence of epileptiform activity. Up to 30 percent of individuals on the spectrum have comorbid epilepsy. These features are beyond the purpose of this review, but are considered in detail in several excellent reviews on the topic [1-4]. Here, we concern ourselves with non-epileptiform activity in the spontaneous EEG and MEG record.

Several studies have examined oscillatory activity in the EEG or MEG in ASD. In an early study, Cantor et al. [124] reported increased delta-band activity (0-4 Hz) in children with ASD compared to mental age-matched control subjects. Frontal abnormalities in slow wave activity have also been suggested by findings of increased frontal theta [125] as well as additional studies replicating increased delta in ASD [126, 127]. There have, however, been opposing reports of reduced delta activity in children with ASD [e.g., see 128].

The alpha rhythm, which is most commonly associated with the posterior-dominant alpha that is reactive to opening and closing the eyes, but which is also observed across multiple regions of cortex reactive to other task-related activities (see below), has also been examined in ASD. In a high density EEG study of resting power and coherence, Murias et al. [127] reported reduced relative alpha power in frontal and occipitoparietal regions in adults with ASD. Increased alpha power was also seen in a recent study, also in an adult ASD sample [129]. Notably, however, in the later study, alpha power was only increased relative to controls in the eyes-open condition and not the eyes-closed condition, whereas the earlier Murias et al. study observed it in the eyes-closed state and did not record eyes-open. A recent MEG paper on eyes-closed resting-state activity in a large sample of children also found elevated alpha in children with ASD, and that elevated alpha power in temporal and parietal regions was positively correlated with scores on the SRS [130]. Although alpha was previously considered to be primarily a cortical idling rhythm [e.g., see 131], more recent evidence suggests that it is associated with top-down, active inhibition of sensory processing in various regions of the cortex in which it is expressed [132].

Spontaneous activity in higher frequency ranges (beta, gamma and higher) has been studied less frequently in ASD. Excessive spontaneous beta-band activity has been reported along midline EEG electrode sites [128] and temporal, parietal and occipital regions in MEG source analyses [130]. Orekhova et al. [133] reported higher levels of EEG gamma-band activity, defined as 24-44 Hz in their study, in two independent samples of boys with ASD (N = 20 each), also reporting that gamma-band activity was correlated with an IQ-derived measure of developmental delay in both samples. No differences were reported in the beta-band. Increased spontaneous gamma has also been reported in independent EEG [134] and MEG studies [130]. Higher frequency oscillatory activity is associated with recurrent inhibitory interneuronal processing in the cerebral cortex [e.g., see 135], which when considered in conjunction with the alpha-band findings, suggests significant and widespread inhibitory dysfunction in the ASD brain.

Beyond simple analyses of spectral power, resting state studies have also examined network connectivity through measures such as coherence. The previously discussed study by Murias et al. [127] also found reduced alpha-band coherence between frontal electrode sites and between frontal and other regions of the brain. Reduced coherence is a common finding in ASD studies [128, 136, 137]. A recent large study involving a retrospective database analysis of 463 cases of ASD and 571 controls found generally reduced coherence across a wide EEG bandwidth (although there were some frequencies that had increased coherence in ASD), and based on spatial patterns was able to achieve an average classification success across age-groups of 88.5% for controls and 86% for ASD subjects.

5. ERSP in ASD

As suggested in section 4, spontaneous oscillatory rhythms have established associations with mental states and are reactive to sensory and motor perturbance. In the following section, this reactivity, the ERSP, is considered for ASD.

5.1. Mu and beta rhythms and mirror neuron dysfunction

One of the most well studied ERSP phenomena in ASD is the mu rhythm ERD, more commonly referred to as mu suppression. The mu rhythm is actually an alpha-band (8-12 Hz, sometimes reported as 8-13 Hz) oscillation, but unlike the more widely known posterior-dominant alpha rhythm, peaks over central sensorimotor regions of the cortex. Intracranial recordings and source localization techniques suggest sources for the mu rhythm in primary sensory and motor and supplementary motor cortices [138, 139]. The

http://dx.doi.org/10.5772/54770

mu rhythm is strongly suppressed preceding movements and maximal suppression occurs contralateral to the movement side [140, 141].

Interest in mu suppression in ASD is focused on the observation that the mu ERD response also occurs when observing the actions of others, not just during the actual motor movements of the subject [142-145]. This has led to a number of studies of mu suppression in the context of mirror neuron theory and the broken mirror hypothesis of ASD [e.g., 146, 147, 148].

EEG findings have been mixed on mu suppression in ASD during action observation. Oberman et al. [146] were first to report that mu suppression was normal to self-performed hand movements, but not to observed movements, in a sample of 10 high-functioning individuals with ASD. A subsequent study by other investigators also reported reduced mu suppression during action observation in 17 adults with ASD [149]. Oberman and colleagues [147] have also presented evidence that mu suppression effects in ASD may depend partly on the familiarity of the observed hand doing the action, as ASD participants' mu suppression only differed from healthy controls when the action was performed by an unfamiliar hand. Two recent EEG studies, however, have reported negative findings between healthy subjects and ASD subjects for the action observation condition. Raymaekers et al. [148] reported equivalent mu suppression action observation in 20 high-functioning children with ASD. Another study found that while ASD participants' imitative actions were impaired behaviorally compared to healthy controls, mu suppression was not impaired to action observation in the same group of 20 individuals [150].

A second sensorimotor rhythm, the beta rhythm that peaks around 20 Hz, has also received attention relating to the broken mirror theory of ASD. The sensorimotor beta rhythm is potentially linked to the mu rhythm as a possible harmonic, but this point is debated because of separate functional localization and timing related to motor activity [e.g., see 151]. This central beta rhythm is suppressed prior to movement onset (ERD) and there is a post-movement power increase (ERS) termed the post-movement beta rebound (PMBR: see Figure 2). Like the mu rhythm, the PMBR has also been shown to be reactive to action observation in addition to actual movement [152]. An early MEG study reported a negative finding with respect to beta activity during action observation, but the lack of finding could have been due to the very small sample (N=5) of ASD participants, or because the focus of the analysis was on the phase-locked evoked beta signal, rather than the induced beta activity that predominates the PMBR [153]. A more recent MEG study, also with a relatively small sample (N=7 ASD subjects) found that the PMBR effect was reduced in the ASD group relative to healthy controls during action observation [154]. Taken together, the EEG and MEG findings on mu and beta rhythm reactivity to action observation appear to suggest that motor regions of the neocortex in ASD may not respond as well to passive observation of another individual's actions, consistent with the prediction of the broken mirror theory in ASD.

5.2. Gamma-band

As with the mu and beta rhythms, interest in the gamma-band has also been keen in ASD. Gamma-band is typically defined as activity in the frequency range between 30 to 80 Hz, although some distinguish between a low (30-80 Hz) and high (> 80 Hz) gamma. This frequency band has generated attention in ASD research primarily because there may be a role for it in cognitive phenomena such as perceptual binding [155-157]. The mechanisms of gamma frequency generation in the cortex and hippocampus are also relatively well characterized [135, 158-160], representing the interaction of pyramidal glutamatergic inputs to fast-spiking interneurons, which in turn recurrently inhibit the pyramidal cells via GABAergic inputs. It is noteworthy that reduced numbers of one of the major classes of these interneurons has been a common finding across animal models of ASD [161].

Grice et al. [82] first examined EEG ERSP changes in gamma-band activity in a face discrimination task. Significant increases in gamma-band power were observed to upright faces than to inverted faces in healthy subjects, but no modulation of gamma power was seen to inverted versus upright faces in the ASD group. A recent MEG paper reported reduced gamma-band phase locking to Mooney face stimuli in ASD, but group differences were only seen when faces were upright, rather than scrambled and inverted [162]. Brown et al. [163] reported increased EEG induced gamma-band power produced in response to perceptual closure of Kanizsa illusory shapes in adolescents with ASD, in contrast to healthy control children who exhibited reduced gamma-band power to the same stimuli. Stroganova et al. [164] also used illusory contour Kanizsa stimuli to demonstrate an absence of early, phase-locked gamma-band responses in occipital electrodes in boys with ASD. A study of visual responses to Gabor patches by reported that children with ASD had attenuated increases in stimulus-induced gamma-band power to increasing spatial frequencies [165], which together with behavioral data was interpreted as evidence for enhanced low-level visual perceptual ability in ASD.

Auditory stimuli produce two types of gamma-band responses; an early, obligatory transient gamma-band response (tGBR) is seen in typically developing individuals to all types of sound stimuli within the first 30-80 ms post-stimulus [166]. When stimuli are modulated in amplitude, either as part of a train of clicks or by amplitude modulation, a later auditory steady-state response (ASSR, see Figure 2) beginning around 100 ms is produced at or near the frequency of modulation, peaking around 40 Hz modulatory rates [167, 168]. Both types of responses are highly phase-locked in typically developing individuals. Wilson et al. [169] reported reduced ASSR power in children and adolescents with ASD. Reduced tGBR power, and reduced phase-locking, has also been reported in both children [42] and adults [170] with ASD using MEG. Both tGBR and ASSR responses are also reported to be impaired in unaffected first-degree relatives of persons with ASD, suggesting that the finding may be a heritable biomarker, or endophenotype [170, 171]. There is an association between visual gamma-band activity and the concentration of GABA in the cerebral cortex [172-174], which as previously discussed may implicate inhibitory dysfunction in the disorder.

6. Future directions

Based on the rich findings emerging in the electrophysiology of autism, there has been an exciting push to expand the field beyond electromagnetic phenomenology and into clinical translation. Two related areas of significant interest have emerged. First, researchers have made significant progress towards successful EEG and MEG recordings in young children and infants so that studies of risk and diagnostic outcome can be conducted. For example, EEG studies examining infants at familial high risk (i.e., a sibling with autism) have shown some promise using measures of EEG complexity and machine learning to separate high-risk infants from typically-developing infants [175, 176]. Of course, a necessary and more difficult next step will be predicting the diagnostic outcome of these infants, since the majority will end up unaffected, or at least below the threshold for clinical diagnosis. A second area of positive development in the field concerns how many of the findings have converged on abnormalities in inhibitory function in relatively well-known circuitry [e.g., see 42], which suggests the possibility of using electrophysiology in clinical trials in terms of prediction of treatment response or even as a primary outcome measure related directly to pharmaceutical intervention.

Acknowledgements

Funding associated with the preparation of this manuscript was provided by the National Institutes of Health grants MH082820 and MH015442.

Author details

Kristina L. McFadden and Donald C. Rojas

Department of Psychiatry, University of Colorado Denver Anschutz Medical Campus, USA

References

[1] Levisohn PM. The autism-epilepsy connection. *Epilepsia* 2007; 48 Suppl 9:33-35.

[2] Spence SJ, Schneider MT. The role of epilepsy and epileptiform EEGs in autism spectrum disorders. *Pediatr Res* 2009; 65:599-606.

[3] Tuchman R, Alessandri M, Cuccaro M. Autism spectrum disorders and epilepsy: moving towards a comprehensive approach to treatment. *Brain Dev* 2010; 32:719-730.

[4] Berg AT, Plioplys S. Epilepsy and autism: is there a special relationship? *Epilepsy Behav* 2012; 23:193-198.

[5] Hamalainen M, Hari R, Ilmoniemi RJ, Knuutila J, Lounasmaa OV. Magnetoencephalography - Theory, Instrumentation, and Applications to Noninvasive Studies of the Working Human Brain. *Reviews of Modern Physics* 1993; 65:413-497.

[6] Dunn MA, Gomes H, Gravel J. Mismatch negativity in children with autism and typical development. *J Autism Dev Disord* 2008; 38:52-71.

[7] Luck SJ.*An Introduction to the Event-Related Potential Technique.* Cambridge, MA: MIT Press; 2005.

[8] Duncan CC, Barry RJ, Connolly JF, Fischer C, Michie PT, Naatanen R, Polich J, Reinvang I, Van Petten C. Event-related potentials in clinical research: guidelines for eliciting, recording, and quantifying mismatch negativity, P300, and N400. *Clin Neurophysiol* 2009; 120:1883-1908.

[9] Makeig S, Debener S, Onton J, Delorme A. Mining event-related brain dynamics. *Trends Cogn Sci* 2004; 8:204-210.

[10] Woods DL. The component structure of the N1 wave of the human auditory evoked potential. *Electroencephalogr Clin Neurophysiol Suppl* 1995; 44:102-109.

[11] Naatanen R, Picton T. The N1 wave of the human electric and magnetic response to sound: a review and an analysis of the component structure. *Psychophysiology* 1987; 24:375-425.

[12] Bomba MD, Pang EW. Cortical auditory evoked potentials in autism: a review. *Int J Psychophysiol* 2004; 53:161-169.

[13] Jeste SS, Nelson CA, 3rd. Event related potentials in the understanding of autism spectrum disorders: an analytical review. *J Autism Dev Disord* 2009; 39:495-510.

[14] Roberts TP, Khan SY, Rey M, Monroe JF, Cannon K, Blaskey L, Woldoff S, Qasmieh S, Gandal M, Schmidt GL, Zarnow DM, Levy SE, Edgar JC. MEG detection of delayed auditory evoked responses in autism spectrum disorders: towards an imaging biomarker for autism. *Autism Res* 2010; 3:8-18.

[15] Scherg M, Von Cramon D. Evoked dipole source potentials of the human auditory cortex. *Electroencephalogr Clin Neurophysiol* 1986; 65:344-360.

[16] Scherg M. Fundamentals of dipole source potential analysis. In *Auditory Evoked Magnetic Fields and Electrical Potentials (Advances in Audiology). Volume* 6. Edited by Grandori F, Hoke M, Romani GL. Switzerland: Karger; 1990: 40-69

[17] Tonnquist-Uhlen I, Borg E, Spens KE. Topography of auditory evoked long-latency potentials in normal children, with particular reference to the N1 component. *Electroencephalogr Clin Neurophysiol* 1995; 95:34-41.

[18] Rojas DC, Walker JR, Sheeder JL, Teale PD, Reite ML. Developmental changes in refractoriness of the neuromagnetic M100 in children. *Neuroreport* 1998; 9:1543-1547.

[19] Edgar JC, Huang MX, Weisend MP, Sherwood A, Miller GA, Adler LE, Canive JM. Interpreting abnormality: an EEG and MEG study of P50 and the auditory paired-stimulus paradigm. *Biol Psychol* 2003; 65:1-20.

[20] Ferri R, Elia M, Agarwal N, Lanuzza B, Musumeci SA, Pennisi G. The mismatch negativity and the P3a components of the auditory event-related potentials in autistic low-functioning subjects. *Clin Neurophysiol* 2003; 114:1671-1680.

[21] Courchesne E, Kilman BA, Galambos R, Lincoln AJ. Autism: processing of novel auditory information assessed by event-related brain potentials. *Electroencephalogr Clin Neurophysiol* 1984; 59:238-248.

[22] Erwin R, Van Lancker D, Guthrie D, Schwafel J, Tanguay P, Buchwald JS. P3 responses to prosodic stimuli in adult autistic subjects. *Electroencephalogr Clin Neurophysiol* 1991; 80:561-571.

[23] Larson MJ, South M, Krauskopf E, Clawson A, Crowley MJ. Feedback and reward processing in high-functioning autism. *Psychiatry Res* 2011; 187:198-203.

[24] Kemner C, Verbaten MN, Cuperus JM, Camfferman G, van Engeland H. Auditory event-related brain potentials in autistic children and three different control groups. *Biol Psychiatry* 1995; 38:150-165.

[25] Martineau J, Garreau B, Barthelemy C, Lelord G. Evoked potentials and P300 during sensory conditioning in autistic children. *Ann N Y Acad Sci* 1984; 425:362-369.

[26] Tecchio F, Benassi F, Zappasodi F, Gialloreti LE, Palermo M, Seri S, Rossini PM. Auditory sensory processing in autism: a magnetoencephalographic study. *Biol Psychiatry* 2003; 54:647-654.

[27] Novick B, Vaughan HG, Jr., Kurtzberg D, Simson R. An electrophysiologic indication of auditory processing defects in autism. *Psychiatry Res* 1980; 3:107-114.

[28] Dunn M, Vaughan HG, Jr., Kreuzer J, Kurtzberg D. Electrophysiologic correlates of semantic classification in autistic and normal children. *Developmental Neuropsychology* 1999; 16:79-99.

[29] Courchesne E, Lincoln AJ, Kilman BA, Galambos R. Event-related brain potential correlates of the processing of novel visual and auditory information in autism. *J Autism Dev Disord* 1985; 15:55-76.

[30] Strandburg RJ, Marsh JT, Brown WS, Asarnow RF, Guthrie D, Higa J. Event-related potentials in high-functioning adult autistics: linguistic and nonlinguistic visual information processing tasks. *Neuropsychologia* 1993; 31:413-434.

[31] Oades RD, Walker MK, Geffen LB, Stern LM. Event-related potentials in autistic and healthy children on an auditory choice reaction time task. *Int J Psychophysiol* 1988; 6:25-37.

[32] Ciesielski KT, Knight JE, Prince RJ, Harris RJ, Handmaker SD. Event-related potentials in cross-modal divided attention in autism. *Neuropsychologia* 1995; 33:225-246.

[33] Townsend J, Westerfield M, Leaver E, Makeig S, Jung T, Pierce K, Courchesne E. Event-related brain response abnormalities in autism: evidence for impaired cerebello-frontal spatial attention networks. *Brain Res Cogn Brain Res* 2001; 11:127-145.

[34] Ceponiene R, Lepisto T, Shestakova A, Vanhala R, Alku P, Naatanen R, Yaguchi K. Speech-sound-selective auditory impairment in children with autism: they can perceive but do not attend. *Proc Natl Acad Sci U S A* 2003; 100:5567-5572.

[35] Hillyard SA. Electrical and magnetic brain recordings: contributions to cognitive neuroscience. *Curr Opin Neurobiol* 1993; 3:217-224.

[36] Woldorff MG, Gallen CC, Hampson SA, Hillyard SA, Pantev C, Sobel D, Bloom FE. Modulation of early sensory processing in human auditory cortex during auditory selective attention. *Proc Natl Acad Sci U S A* 1993; 90:8722-8726.

[37] Lincoln AJ, Courchesne E, Harms L, Allen M. Contextual probability evaluation in autistic, receptive developmental language disorder, and control children: event-related brain potential evidence. *J Autism Dev Disord* 1993; 23:37-58.

[38] Oram Cardy JE, Ferrari P, Flagg EJ, Roberts W, Roberts TP. Prominence of M50 auditory evoked response over M100 in childhood and autism. *Neuroreport* 2004; 15:1867-1870.

[39] Sokhadze E, Baruth J, Tasman A, Sears L, Mathai G, El-Baz A, Casanova MF. Event-related potential study of novelty processing abnormalities in autism. *Appl Psychophysiol Biofeedback* 2009; 34:37-51.

[40] Gage NM, Siegel B, Roberts TP. Cortical auditory system maturational abnormalities in children with autism disorder: an MEG investigation. *Brain Res Dev Brain Res* 2003; 144:201-209.

[41] Bruneau N, Roux S, Adrien JL, Barthelemy C. Auditory associative cortex dysfunction in children with autism: evidence from late auditory evoked potentials (N1 wave-T complex). *Clin Neurophysiol* 1999; 110:1927-1934.

[42] Gandal MJ, Edgar JC, Ehrlichman RS, Mehta M, Roberts TP, Siegel SJ. Validating gamma oscillations and delayed auditory responses as translational biomarkers of autism. *Biol Psychiatry* 2010; 68:1100-1106.

[43] Siegal M, Blades M. Language and auditory processing in autism. *Trends Cogn Sci* 2003; 7:378-380.

[44] Rapin I, Dunn M. Update on the language disorders of individuals on the autistic spectrum. *Brain Dev* 2003; 25:166-172.

[45] Gomot M, Giard MH, Adrien JL, Barthelemy C, Bruneau N. Hypersensitivity to acoustic change in children with autism: electrophysiological evidence of left frontal cortex dysfunctioning. *Psychophysiology* 2002; 39:577-584.

[46] Bruneau N, Roux S, Guerin P, Barthelemy C, Lelord G. Temporal prominence of auditory evoked potentials (N1 wave) in 4-8-year-old children. *Psychophysiology* 1997; 34:32-38.

[47] Roberts TP, Schmidt GL, Egeth M, Blaskey L, Rey MM, Edgar JC, Levy SE. Electrophysiological signatures: magnetoencephalographic studies of the neural correlates of language impairment in autism spectrum disorders. *Int J Psychophysiol* 2008; 68:149-160.

[48] Schmidt GL, Rey MM, Oram Cardy JE, Roberts TP. Absence of M100 source asymmetry in autism associated with language functioning. *Neuroreport* 2009; 20:1037-1041.

[49] Elberling C, Bak C, Kofoed B, Lebech J, Saermark K. Auditory magnetic fields from the human cerebral cortex: location and strength of an equivalent current dipole. *Acta Neurol Scand* 1982; 65:553-569.

[50] American Psychiatric Association *Diagnostic and Statistical Manual of Mental Disorders.* 4th edn. Washington, D.C.: American Psychiatric Association; (1994).

[51] Roberts TP, Heiken K, Kahn SY, Qasmieh S, Blaskey L, Solot C, Parker WA, Verma R, Edgar JC. Delayed magnetic mismatch negativity field, but not auditory M100 response, in specific language impairment. *Neuroreport* 2012; 23:463-468.

[52] Naatanen R. The mismatch negativity: a powerful tool for cognitive neuroscience. *Ear Hear* 1995; 16:6-18.

[53] Naatanen R, Paavilainen P, Rinne T, Alho K. The mismatch negativity (MMN) in basic research of central auditory processing: a review. *Clin Neurophysiol* 2007; 118:2544-2590.

[54] Naatanen R, Alho K. Mismatch negativity--a unique measure of sensory processing in audition. *Int J Neurosci* 1995; 80:317-337.

[55] Naatanen R, Kujala T, Escera C, Baldeweg T, Kreegipuu K, Carlson S, Ponton C. The mismatch negativity (MMN)--a unique window to disturbed central auditory processing in ageing and different clinical conditions. *Clin Neurophysiol* 2012; 123:424-458.

[56] Lepisto T, Kajander M, Vanhala R, Alku P, Huotilainen M, Naatanen R, Kujala T. The perception of invariant speech features in children with autism. *Biol Psychol* 2008; 77:25-31.

[57] Bodatsch M, Ruhrmann S, Wagner M, Muller R, Schultze-Lutter F, Frommann I, Brinkmeyer J, Gaebel W, Maier W, Klosterkotter J, Brockhaus-Dumke A. Prediction of psychosis by mismatch negativity. *Biol Psychiatry* 2011; 69:959-966.

[58] Winkler I, Kushnerenko E, Horvath J, Ceponiene R, Fellman V, Huotilainen M, Naatanen R, Sussman E. Newborn infants can organize the auditory world. *Proc Natl Acad Sci U S A* 2003; 100:11812-11815.

[59] Lepisto T, Kujala T, Vanhala R, Alku P, Huotilainen M, Naatanen R. The discrimination of and orienting to speech and non-speech sounds in children with autism. *Brain Res* 2005; 1066:147-157.

[60] Kasai K, Hashimoto O, Kawakubo Y, Yumoto M, Kamio S, Itoh K, Koshida I, Iwanami A, Nakagome K, Fukuda M, Yamasue H, Yamada H, Abe O, Aoki S, Kato N. Delayed automatic detection of change in speech sounds in adults with autism: a magnetoencephalographic study. *Clin Neurophysiol* 2005; 116:1655-1664.

[61] Oram Cardy JE, Flagg EJ, Roberts W, Roberts TP. Delayed mismatch field for speech and non-speech sounds in children with autism. *Neuroreport* 2005; 16:521-525.

[62] Roberts TP, Cannon KM, Tavabi K, Blaskey L, Khan SY, Monroe JF, Qasmieh S, Levy SE, Edgar JC. Auditory magnetic mismatch field latency: a biomarker for language impairment in autism. *Biol Psychiatry* 2011; 70:263-269.

[63] Woldorff MG, Hackley SA, Hillyard SA. The effects of channel-selective attention on the mismatch negativity wave elicited by deviant tones. *Psychophysiology* 1991; 28:30-42.

[64] Naatanen R, Paavilainen P, Tiitinen H, Jiang D, Alho K. Attention and mismatch negativity. *Psychophysiology* 1993; 30:436-450.

[65] Bonnel A, Mottron L, Peretz I, Trudel M, Gallun E, Bonnel AM. Enhanced pitch sensitivity in individuals with autism: a signal detection analysis. *J Cogn Neurosci* 2003; 15:226-235.

[66] Bentin S, Allison T, Puce A, Perez E, McCarthy G. Electrophysiological Studies of Face Perception in Humans. *J Cogn Neurosci* 1996; 8:551-565.

[67] Jemel B, Mottron L, Dawson M. Impaired face processing in autism: fact or artifact? *J Autism Dev Disord* 2006; 36:91-106.

[68] Webb SJ, Merkle K, Murias M, Richards T, Aylward E, Dawson G. ERP responses differentiate inverted but not upright face processing in adults with ASD. *Soc Cogn Affect Neurosci* 2012; 7:578-587.

[69] Churches O, Damiano C, Baron-Cohen S, Ring H. Getting to know you: the acquisition of new face representations in autism spectrum conditions. *Neuroreport* 2012; 23:668-672.

[70] Batty M, Meaux E, Wittemeyer K, Roge B, Taylor MJ. Early processing of emotional faces in children with autism: An event-related potential study. *J Exp Child Psychol* 2011; 109:430-444.

[71] Bentin S, Deouell LY. Structural encoding and identification in face processing: erp evidence for separate mechanisms. *Cogn Neuropsychol* 2000; 17:35-55.

[72] McPartland J, Dawson G, Webb SJ, Panagiotides H, Carver LJ. Event-related brain potentials reveal anomalies in temporal processing of faces in autism spectrum disorder. *J Child Psychol Psychiatry* 2004; 45:1235-1245.

[73] Eimer M. Effects of face inversion on the structural encoding and recognition of faces. Evidence from event-related brain potentials. *Brain Res Cogn Brain Res* 2000; 10:145-158.

[74] Rossion B, Gauthier I, Tarr MJ, Despland P, Bruyer R, Linotte S, Crommelinck M. The N170 occipito-temporal component is delayed and enhanced to inverted faces but not to inverted objects: an electrophysiological account of face-specific processes in the human brain. *Neuroreport* 2000; 11:69-74.

[75] Grice SJ, Halit H, Farroni T, Baron-Cohen S, Bolton P, Johnson MH. Neural correlates of eye-gaze detection in young children with autism. *Cortex* 2005; 41:342-353.

[76] Dawson G, Webb SJ, McPartland J. Understanding the nature of face processing impairment in autism: insights from behavioral and electrophysiological studies. *Dev Neuropsychol* 2005; 27:403-424.

[77] Schweinberger SR, Kaufmann JM, Moratti S, Keil A, Burton AM. Brain responses to repetitions of human and animal faces, inverted faces, and objects: an MEG study. *Brain Res* 2007; 1184:226-233.

[78] Carey S. Becoming a face expert. *Philos Trans R Soc Lond B Biol Sci* 1992; 335:95-102; discussion 102-103.

[79] Tanaka JW, Curran T. A neural basis for expert object recognition. *Psychol Sci* 2001; 12:43-47.

[80] Hileman CM, Henderson H, Mundy P, Newell L, Jaime M. Developmental and individual differences on the P1 and N170 ERP components in children with and without autism. *Dev Neuropsychol* 2011; 36:214-236.

[81] Churches O, Baron-Cohen S, Ring H. The psychophysiology of narrower face processing in autism spectrum conditions. *Neuroreport* 2012; 23:395-399.

[82] Grice SJ, Spratling MW, Karmiloff-Smith A, Halit H, Csibra G, de Haan M, Johnson MH. Disordered visual processing and oscillatory brain activity in autism and Williams syndrome. *Neuroreport* 2001; 12:2697-2700.

[83] O'Connor K, Hamm JP, Kirk IJ. The neurophysiological correlates of face processing in adults and children with Asperger's syndrome. *Brain Cogn* 2005; 59:82-95.

[84] O'Connor K, Hamm JP, Kirk IJ. Neurophysiological responses to face, facial regions and objects in adults with Asperger's syndrome: an ERP investigation. *Int J Psychophysiol* 2007; 63:283-293.

[85] Dawson G, Carver L, Meltzoff AN, Panagiotides H, McPartland J, Webb SJ. Neural correlates of face and object recognition in young children with autism spectrum disorder, developmental delay, and typical development. *Child Dev* 2002; 73:700-717.

[86] Joseph RM, Tanaka J. Holistic and part-based face recognition in children with autism. *Journal of Child Psychology and Psychiatry and Allied Disciplines* 2003; 44:529-542.

[87] Churches O, Wheelwright S, Baron-Cohen S, Ring H. The N170 is not modulated by attention in autism spectrum conditions. *Neuroreport* 2010; 21:399-403.

[88] Bailey AJ, Braeutigam S, Jousmaki V, Swithenby SJ. Abnormal activation of face processing systems at early and intermediate latency in individuals with autism spectrum disorder: a magnetoencephalographic study. *Eur J Neurosci* 2005; 21:2575-2585.

[89] Kanner L. Early infantile autism revisited. *Psychiatry Dig* 1968; 29:17-28.

[90] Senju A, Johnson MH. Atypical eye contact in autism: models, mechanisms and development. *Neurosci Biobehav Rev* 2009; 33:1204-1214.

[91] Baranek GT. Autism during infancy: a retrospective video analysis of sensory-motor and social behaviors at 9-12 months of age. *J Autism Dev Disord* 1999; 29:213-224.

[92] Clifford S, Young R, Williamson P. Assessing the early characteristics of autistic disorder using video analysis. *J Autism Dev Disord* 2007; 37:301-313.

[93] Osterling J, Dawson G. Early recognition of children with autism: a study of first birthday home videotapes. *J Autism Dev Disord* 1994; 24:247-257.

[94] Behrmann M, Thomas C, Humphreys K. Seeing it differently: visual processing in autism. *Trends Cogn Sci* 2006; 10:258-264.

[95] Taylor MJ, Batty M, Itier RJ. The faces of development: a review of early face processing over childhood. *J Cogn Neurosci* 2004; 16:1426-1442.

[96] Picton TW. The P300 wave of the human event-related potential. *J Clin Neurophysiol* 1992; 9:456-479.

[97] Polich J. Updating P300: an integrative theory of P3a and P3b. *Clin Neurophysiol* 2007; 118:2128-2148.

[98] Marco EJ, Hinkley LB, Hill SS, Nagarajan SS. Sensory processing in autism: a review of neurophysiologic findings. *Pediatr Res* 2011; 69:48R-54R.

[99] Polich J, Kok A. Cognitive and biological determinants of P300: an integrative review. *Biol Psychol* 1995; 41:103-146.

[100] Gomot M, Belmonte MK, Bullmore ET, Bernard FA, Baron-Cohen S. Brain hyper-reactivity to auditory novel targets in children with high-functioning autism. *Brain* 2008; 131:2479-2488.

[101] Dawson G, Finley C, Phillips S, Galpert L, Lewy A. Reduced P3 amplitude of the event-related brain potential: its relationship to language ability in autism. *J Autism Dev Disord* 1988; 18:493-504.

[102] Courchesne E, Lincoln AJ, Yeung-Courchesne R, Elmasian R, Grillon C. Pathophysiologic findings in nonretarded autism and receptive developmental language disorder. *J Autism Dev Disord* 1989; 19:1-17.

[103] Ciesielski KT, Courchesne E, Elmasian R. Effects of focused selective attention tasks on event-related potentials in autistic and normal individuals. *Electroencephalogr Clin Neurophysiol* 1990; 75:207-220.

[104] Verbaten MN, Roelofs JW, van Engeland H, Kenemans JK, Slangen JL. Abnormal visual event-related potentials of autistic children. *J Autism Dev Disord* 1991; 21:449-470.

[105] Pritchard WS, Raz N, August GJ. Visual augmenting/reducing and P300 in autistic children. *J Autism Dev Disord* 1987; 17:231-242.

[106] Salmond CH, Vargha-Khademl F, Gadian DG, de Haan M, Baldeweg T. Heterogeneity in the patterns of neural abnormality in autistic spectrum disorders: evidence from ERP and MRI. *Cortex* 2007; 43:686-699.

[107] Whitehouse AJ, Barry JG, Bishop DV. Further defining the language impairment of autism: is there a specific language impairment subtype? *J Commun Disord* 2008; 41:319-336.

[108] Kutas M, Hillyard SA. Reading senseless sentences: brain potentials reflect semantic incongruity. *Science* 1980; 207:203-205.

[109] Osterhout L, McLaughlin J, Bersick M. Event-related brain potentials and human language. *Trends Cogn Sci* 1997; 1:203-209.

[110] van Berkum JJA, Brown CM, Hagoort P. Early referential context effects in sentence processing: Evidence from event-related brain potentials. *Journal of Memory and Language* 1999; 41:147-182.

[111] Kutas M, Federmeier KD. Electrophysiology reveals semantic memory use in language comprehension. *Trends Cogn Sci* 2000; 4:463-470.

[112] Bentin S, McCarthy G, Wood CC. Event-related potentials, lexical decision and semantic priming. *Electroencephalogr Clin Neurophysiol* 1985; 60:343-355.

[113] Fishman I, Yam A, Bellugi U, Lincoln A, Mills D. Contrasting patterns of language-associated brain activity in autism and Williams syndrome. *Soc Cogn Affect Neurosci* 2011; 6:630-638.

[114] Henson RN. Neuroimaging studies of priming. *Prog Neurobiol* 2003; 70:53-81.

[115] Schacter DL, Wig GS, Stevens WD. Reductions in cortical activity during priming. *Curr Opin Neurobiol* 2007; 17:171-176.

[116] Brown C, Hagoort P. The Processing Nature of the N400 - Evidence from Masked Priming. *Journal of Cognitive Neuroscience* 1993; 5:34-44.

[117] Pijnacker J, Geurts B, van Lambalgen M, Buitelaar J, Hagoort P. Exceptions and anomalies: an ERP study on context sensitivity in autism. *Neuropsychologia* 2010; 48:2940-2951.

[118] Dunn MA, Bates JC. Developmental change in neutral processing of words by children with autism. *J Autism Dev Disord* 2005; 35:361-376.

[119] McCleery JP, Ceponiene R, Burner KM, Townsend J, Kinnear M, Schreibman L. Neural correlates of verbal and nonverbal semantic integration in children with autism spectrum disorders. *J Child Psychol Psychiatry* 2010; 51:277-286.

[120] Henderson LM, Clarke PJ, Snowling MJ. Accessing and selecting word meaning in autism spectrum disorder. *J Child Psychol Psychiatry* 2011; 52:964-973.

[121] Jolliffe T, Baron-Cohen S. A test of central coherence theory: linguistic processing in high-functioning adults with autism or Asperger syndrome: is local coherence impaired? *Cognition* 1999; 71:149-185.

[122] Jolliffe T, Baron-Cohen S. Linguistic processing in high-functioning adults with autism or Asperger's syndrome. Is global coherence impaired? *Psychol Med* 2000; 30:1169-1187.

[123] Braeutigam S, Swithenby SJ, Bailey AJ. Contextual integration the unusual way: a magnetoencephalographic study of responses to semantic violation in individuals with autism spectrum disorders. *Eur J Neurosci* 2008; 27:1026-1036.

[124] Cantor DS, Thatcher RW, Hrybyk M, Kaye H. Computerized EEG analyses of autistic children. *J Autism Dev Disord* 1986; 16:169-187.

[125] Daoust AM, Limoges E, Bolduc C, Mottron L, Godbout R. EEG spectral analysis of wakefulness and REM sleep in high functioning autistic spectrum disorders. *Clin Neurophysiol* 2004; 115:1368-1373.

[126] Chan AS, Sze SL, Cheung MC. Quantitative electroencephalographic profiles for children with autistic spectrum disorder. *Neuropsychology* 2007; 21:74-81.

[127] Murias M, Webb SJ, Greenson J, Dawson G. Resting state cortical connectivity reflected in EEG coherence in individuals with autism. *Biol Psychiatry* 2007; 62:270-273.

[128] Coben R, Clarke AR, Hudspeth W, Barry RJ. EEG power and coherence in autistic spectrum disorder. *Clin Neurophysiol* 2008; 119:1002-1009.

[129] Mathewson KJ, Jetha MK, Drmic IE, Bryson SE, Goldberg JO, Schmidt LA. Regional EEG alpha power, coherence, and behavioral symptomatology in autism spectrum disorder. *Clin Neurophysiol* 2012; 123:1798-1809.

[130] Cornew L, Roberts TP, Blaskey L, Edgar JC. Resting-state oscillatory activity in autism spectrum disorders. *J Autism Dev Disord* 2012; 42:1884-1894.

[131] Pfurtscheller G, Stancak A, Jr., Neuper C. Event-related synchronization (ERS) in the alpha band--an electrophysiological correlate of cortical idling: a review. *Int J Psychophysiol* 1996; 24:39-46.

[132] Klimesch W, Sauseng P, Hanslmayr S. EEG alpha oscillations: the inhibition-timing hypothesis. *Brain Res Rev* 2007; 53:63-88.

[133] Orekhova EV, Stroganova TA, Nygren G, Tsetlin MM, Posikera IN, Gillberg C, Elam M. Excess of high frequency electroencephalogram oscillations in boys with autism. *Biol Psychiatry* 2007; 62:1022-1029.

[134] Sheikhani A, Behnam H, Noroozian M, Mohammadi MR, Mohammadi M. Abnormalities of quantitative electroencephalography in children with Asperger disorder in various conditions. *Research in Autism Spectrum Disorders* 2009; 3:538-546.

[135] Bartos M, Vida I, Jonas P. Synaptic mechanisms of synchronized gamma oscillations in inhibitory interneuron networks. *Nat Rev Neurosci* 2007; 8:45-56.

[136] Barttfeld P, Wicker B, Cukier S, Navarta S, Lew S, Sigman M. A big-world network in ASD: dynamical connectivity analysis reflects a deficit in long-range connections and an excess of short-range connections. *Neuropsychologia* 2011; 49:254-263.

[137] Duffy FH, Als H. A stable pattern of EEG spectral coherence distinguishes children with autism from neuro-typical controls - a large case control study. *BMC medicine* 2012; 10:64.

[138] Tiihonen J, Kajola M, Hari R. Magnetic mu rhythm in man. *Neuroscience* 1989; 32:793-800.

[139] Miller KJ, Leuthardt EC, Schalk G, Rao RP, Anderson NR, Moran DW, Miller JW, Ojemann JG. Spectral changes in cortical surface potentials during motor movement. *J Neurosci* 2007; 27:2424-2432.

[140] Pfurtscheller G, Neuper C, Andrew C, Edlinger G. Foot and hand area mu rhythms. *Int J Psychophysiol* 1997; 26:121-135.

[141] Babiloni C, Carducci F, Cincotti F, Rossini PM, Neuper C, Pfurtscheller G, Babiloni F. Human movement-related potentials vs desynchronization of EEG alpha rhythm: a high-resolution EEG study. *Neuroimage* 1999; 10:658-665.

[142] Cochin S, Barthelemy C, Roux S, Martineau J. Observation and execution of movement: similarities demonstrated by quantified electroencephalography. *Eur J Neurosci* 1999; 11:1839-1842.

[143] Muthukumaraswamy SD, Johnson BW. Changes in rolandic mu rhythm during observation of a precision grip. *Psychophysiology* 2004; 41:152-156.

[144] Muthukumaraswamy SD, Johnson BW, McNair NA. Mu rhythm modulation during observation of an object-directed grasp. *Brain Res Cogn Brain Res* 2004; 19:195-201.

[145] Hari R, Forss N, Avikainen S, Kirveskari E, Salenius S, Rizzolatti G. Activation of human primary motor cortex during action observation: a neuromagnetic study. *Proc Natl Acad Sci U S A* 1998; 95:15061-15065.

[146] Oberman LM, Hubbard EM, McCleery JP, Altschuler EL, Ramachandran VS, Pineda JA. EEG evidence for mirror neuron dysfunction in autism spectrum disorders. *Brain Res Cogn Brain Res* 2005; 24:190-198.

[147] Oberman LM, Ramachandran VS, Pineda JA. Modulation of mu suppression in children with autism spectrum disorders in response to familiar or unfamiliar stimuli: the mirror neuron hypothesis. *Neuropsychologia* 2008; 46:1558-1565.

[148] Raymaekers R, Wiersema JR, Roeyers H. EEG study of the mirror neuron system in children with high functioning autism. *Brain Res* 2009; 1304:113-121.

[149] Bernier R, Dawson G, Webb S, Murias M. EEG mu rhythm and imitation impairments in individuals with autism spectrum disorder. *Brain Cogn* 2007; 64:228-237.

[150] Fan YT, Decety J, Yang CY, Liu JL, Cheng Y. Unbroken mirror neurons in autism spectrum disorders. *J Child Psychol Psychiatry* 2010; 51:981-988.

[151] Hari R, Salmelin R. Human cortical oscillations: a neuromagnetic view through the skull. *Trends Neurosci* 1997; 20:44-49.

[152] Muthukumaraswamy SD, Johnson BW. Primary motor cortex activation during action observation revealed by wavelet analysis of the EEG. *Clin Neurophysiol* 2004; 115:1760-1766.

[153] Avikainen S, Kulomaki T, Hari R. Normal movement reading in Asperger subjects. *Neuroreport* 1999; 10:3467-3470.

[154] Honaga E, Ishii R, Kurimoto R, Canuet L, Ikezawa K, Takahashi H, Nakahachi T, Iwase M, Mizuta I, Yoshimine T, Takeda M. Post-movement beta rebound abnormality as indicator of mirror neuron system dysfunction in autistic spectrum disorder: an MEG study. *Neurosci Lett* 2010; 478:141-145.

[155] Freeman WJ.*Mass Action in the Nervous System.* New York: Academic Press; 1975.

[156] Belmonte MK, Allen G, Beckel-Mitchener A, Boulanger LM, Carper RA, Webb SJ. Autism and abnormal development of brain connectivity. *J Neurosci* 2004; 24:9228-9231.

[157] Uhlhaas PJ, Singer W. Neural synchrony in brain disorders: relevance for cognitive dysfunctions and pathophysiology. *Neuron* 2006; 52:155-168.

[158] Gloveli T, Dugladze T, Saha S, Monyer H, Heinemann U, Traub RD, Whittington MA, Buhl EH. Differential involvement of oriens/pyramidale interneurones in hippocampal network oscillations in vitro. *J Physiol* 2005; 562:131-147.

[159] Whittington MA, Traub RD, Jefferys JG. Synchronized oscillations in interneuron networks driven by metabotropic glutamate receptor activation. *Nature* 1995; 373:612-615.

[160] Hajos N, Palhalmi J, Mann EO, Nemeth B, Paulsen O, Freund TF. Spike timing of distinct types of GABAergic interneuron during hippocampal gamma oscillations in vitro. *J Neurosci* 2004; 24:9127-9137.

[161] Gogolla N, Leblanc JJ, Quast KB, Sudhof TC, Fagiolini M, Hensch TK. Common circuit defect of excitatory-inhibitory balance in mouse models of autism. *Journal of neurodevelopmental disorders* 2009; 1:172-181.

[162] Sun L, Grutzner C, Bolte S, Wibral M, Tozman T, Schlitt S, Poustka F, Singer W, Freitag CM, Uhlhaas PJ. Impaired Gamma-Band Activity during Perceptual Organization in Adults with Autism Spectrum Disorders: Evidence for Dysfunctional Network Activity in Frontal-Posterior Cortices. *The Journal of neuroscience : the official journal of the Society for Neuroscience* 2012; 32:9563-9573.

[163] Brown C, Gruber T, Boucher J, Rippon G, Brock J. Gamma abnormalities during perception of illusory figures in autism. *Cortex* 2005; 41:364-376.

[164] Stroganova TA, Orekhova EV, Prokofyev AO, Tsetlin MM, Gratchev VV, Morozov AA, Obukhov YV. High-frequency oscillatory response to illusory contour in typically developing boys and boys with autism spectrum disorders. *Cortex; a journal devoted to the study of the nervous system and behavior* 2012; 48:701-717.

[165] Milne E, Scope A, Pascalis O, Buckley D, Makeig S. Independent component analysis reveals atypical electroencephalographic activity during visual perception in individuals with autism. *Biological psychiatry* 2009; 65:22-30.

[166] Pantev C, Makeig S, Hoke M, Galambos R, Hampson S, Gallen C. Human auditory evoked gamma-band magnetic fields. *Proc Natl Acad Sci U S A* 1991; 88:8996-9000.

[167] Azzena GB, Conti G, Santarelli R, Ottaviani F, Paludetti G, Maurizi M. Generation of human auditory steady-state responses (SSRs). I: Stimulus rate effects. *Hear Res* 1995; 83:1-8.

[168] Hari R, Hamalainen M, Joutsiniemi SL. Neuromagnetic steady-state responses to auditory stimuli. *J Acoust Soc Am* 1989; 86:1033-1039.

[169] Wilson TW, Rojas DC, Reite ML, Teale PD, Rogers SJ. Children and adolescents with autism exhibit reduced MEG steady-state gamma responses. *Biol Psychiatry* 2007; 62:192-197.

[170] Rojas DC, Maharajh K, Teale P, Rogers SJ. Reduced neural synchronization of gamma-band MEG oscillations in first-degree relatives of children with autism. *BMC Psychiatry* 2008; 8:66.

[171] Rojas DC, Teale PD, Maharajh K, Kronberg E, Youngpeter K, Wilson L, Wallace A, Hepburn S. Transient and steady-state auditory gamma-band responses in first-degree relatives of people with autism spectrum disorder. *Mol Autism* 2011; 2:11.

[172] Muthukumaraswamy SD, Singh KD, Swettenham JB, Jones DK. Visual gamma oscillations and evoked responses: variability, repeatability and structural MRI correlates. *Neuroimage* 2010; 49:3349-3357.

[173] Muthukumaraswamy SD, Edden RA, Jones DK, Swettenham JB, Singh KD. Resting GABA concentration predicts peak gamma frequency and fMRI amplitude in response to visual stimulation in humans. *Proc Natl Acad Sci U S A* 2009; 106:8356-8361.

[174] Edden RA, Muthukumaraswamy SD, Freeman TC, Singh KD. Orientation discrimination performance is predicted by GABA concentration and gamma oscillation frequency in human primary visual cortex. *J Neurosci* 2009; 29:15721-15726.

[175] Bosl W, Tierney A, Tager-Flusberg H, Nelson C. EEG complexity as a biomarker for autism spectrum disorder risk. *BMC medicine* 2011; 9:18.

[176] Stahl D, Pickles A, Elsabbagh M, Johnson MH. Novel machine learning methods for ERP analysis: a validation from research on infants at risk for autism. *Developmental Neuropsychology* 2012; 37:274-298.

[177] McFadden, K.L., Healy, K.M., Dettmann, M.L., Kaye, J.T., Ito, T.A., Hernandez, T.D. (2011). Acupressure as a non-pharmacological intervention for traumatic brain injury (TBI). Journal of neurotrauma 28, 21-34.

Section 4

Treatment

Chapter 5

The Sensory Experience of Toilet Training and Its Implications for Autism Intervention

Jane Yip, Betsy Powers and Fengyi Kuo

Additional information is available at the end of the chapter

http://dx.doi.org/10.5772/54947

1. Introduction

Individuals with autism spectrum disorders are often plagued with incontinence due to compromised sensory processing between the peripheral and central nervous system. Currently, a combination of the Azrin & Foxx method (1971) and operant conditioning are considered the standard protocol for toilet training children with developmental and intellectual disabilities including individuals with autism. Although programs that have been adapted to children with physical disabilities have resulted in successful toilet training in most cases, there is a considerable proportion of individuals with autism who have reached adulthood without being accomplished in bladder and bowel control. These individuals often reside in the spectrum of non-verbal with lower cognition levels (Dalrymple & Ruble, 1992) and exhibit not only incontinence but are challenged with abnormality in toileting such as encopresis, enuresis and fecal smearing. A lack of neurobiological data precluded conclusions regarding the development of, and abnormality in, toilet training individuals with autism. Existing literature is skewed towards treating the aforementioned issue as solely the domain of behavior. Whereas toilet training is both an inherently neurobiological and cultural phenomenon, society's failure to comprehend and address the problem of some of our most vulnerable members in the autism spectrum reflect a still "Victorian-albeit taboo" in this fundamentally important issue in self-care. This chapter reviews the importance of independent self-care skills in family routines, and proposes a putative sensory-neural understanding of the toilet training-resistant cohort, which is largely unknown. Case studies of two children with autism demonstrate how sensory experiences may affect toilet training success.

2. Toilet training in autism

Toilet training is commonly delayed in Autism spectrum disorders (ASD). Dalrymple and Ruble (1992) surveyed 100 parents of clients with ASD (mean age = 19.5 years) and reported that 22% of them did not have full success with toileting. Research studies on toileting in autism have been scarce and are mainly single case studies (Kroeger & Sorensen, 2010). In general, toilet training in autism included: the teaching of hand-washing techniques and cleaning up strategies (Rumfelt-Wright, 2001; Furman, 2001), positive motivational practice, habituation to the bathroom (Cicero & Pfadt, 2002; Rockville, 2008) and the utilization of communication training component to promote self-initiated toileting (LeBlanc, 2005).

3. Impact of lack of independence in toileting in family's life

Children with ASD exhibiting sensory processing issues often lack bladder and bowel control. This can impact family routines. Most notably, the families of children with ASD tended to avoid social participation more so than families with typically developing children (Bagby et al., 2012). Gray (1994) reports that social withdrawal is often a coping strategy for families with children on the autism spectrum who display inappropriate toileting behavior. Additionally, mothers of children with autism are less likely to work full-time than mothers of typically developing children (Porterfield, 2004). As schools and child-care facilities often reject children who are not independent with toileting, parental participation in activities outside of the homes may be further limited as a result of caregiving roles and extensive responsibilities.

4. Self-care skills in children with ASD: A brief overview

Most child care centers and inclusive preschools require children to be toilet trained before entering a certain grade level. For example, inclusive preschools in which prompted independent urination could occur during group lavatory trips at structured times. Children with ASD whom require diaper changes during the school days would be excluded from this classroom routine. As a longer period of seated time is expected for learning to occur, the children attending inclusive schools are required to maintain sufficient bladder and bowel control and to sustain basic self-care needs in order to adhere to environmental structures. Therefore, the ability of children to learn and refine performance of basic self-care tasks holds meaning not only for the children but also for their caregivers and teachers.

5. Neurobiology of toilet training

Without understanding how the body orchestrate the many events associated with going to the bathroom, it is not possible to progress in the field of intervention for individuals still

http://dx.doi.org/10.5772/54947

struggling with containing themselves appropriately past the critical period for toilet training. This is because toilet training is a behavior that lies in the juncture between a biology that is involuntary and a biology governed by free will that is voluntary. In neurobiological terms, conscious control of urinary and anal continence requires an intact communication between the peripheral and central nervous system. Therefore, dysfunction of the nervous system would result in the loss of control of continence and could explain the prevalence of incontinence in individuals with compromised neurology such as autism. Micturition or the urge to urinate is dependent on the integrity of a center in the lumbo-sacral region of the cord with the bladder and urethra. Micturition takes place when the bladder and urethra sphincter hence stimulated transmits impulses via the ganglia to the cord. Whereas micturition can take place with the lower end of the cord alone, the voluntary control of micturition requires coordination with the central nervous system. Micturition reflex was evoked by filling the bladder and is mediated through sacral spinal region as illustrated in Figure 1. The micturition reflex arch involves the coordination from the spinal cord to the brainstem-cerebellum and back (Fowler et al., 2008). The micturition reflex arch composed of an afferent pathway from the urinary bladder to the lumbar and sacral region onto the spinal cord and to the pontine micturition center before connecting to higher centers in the brain. The afferent pathway is served by the pelvic nerve and hypogastric nerve which connects to the spinal cord before entering the brain. The efferent pathway, on the other hand, projects from the higher center of the brain through the periaqueductal gray area (PAG), travelling to the bladder, then to the lumbro-sacral parasympathetic center and finally to the sphincter muscle of the bladder via the pudendal nerve (Manto & Jissendi, 2012; see Figure 1). The pontine micturition center is thought to be the control center for micturition and is involuntary. The instinctual and involuntary aspects aside, there is a socio-cultural dimension of micturition and proceeds by conscious decision which develops through training. The voluntary center needs to recruit more elaborate coordination between the higher centers in the brain with the spinal cord. This conscious control of urination is thought to recruit higher cerebral centers including the medial prefrontal cortex, anterior cingulate cortex and insular cortex which will communicate with the pontine micturition center that mastermind toilet training. Figure 2 describes in detail the interplay of multiple systems that are involved in the conscious control of continence. As for for the control anal defecation: it is similar to the micturition reflex and work via the coordination of various anorectal physiology: action of sacral nerve stimulation, neurological control of the colon and anorectum and higher cerebral centers (Gourcerol et al., 2011).

Figure 2 illustrates the multi-level control systems that are required for the conscious control of bladder voiding. The PAG is recruited to connect lower spinal cord to the cerebral cortex. Since the cerebellum forms projections to the PAG via the pons, it is likely that sensory processing entering from the cerebellum convey information to the PAG and ultimately communicates with the cerebral cortex.

The neurobiology of continence control indicates that the frontal cortex and cerebellum is part of the essential circuit during the conscious control of continence. Successful toilet training should point toward the contribution of cognitive function and habituation that recruits the frontal cortex and the sensory components of the cerebellum. This paper strive to understand if the dual combination therapy targeting (i) cognitive and habituation training, and (ii) sensory modulation could result in positive outcomes for individuals resistant to traditional approaches to toilet training.

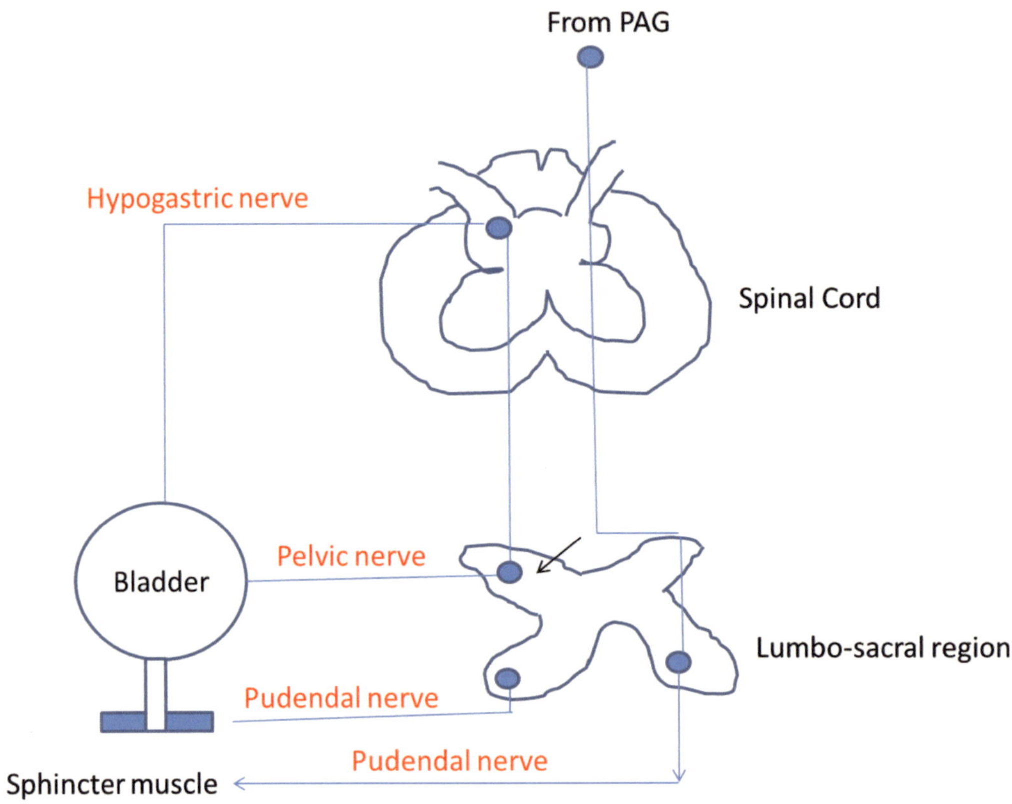

Figure 1. Connections between the afferent pathway (from the bladder to the lumbro-sacral region and the spinal cord via the hypogastric and pelvic nerve) and the efferent pathway (from the brain through the PAG, the spinal cord and onto the sphincter muscle and bladder via the pudendal nerve) that gives rise to the involuntary control of micturition or bladder voiding (Adapted from Manto and Jissendi, 2012).

During the maximal storage of urine, the pressure against the bladder stimulates nervous reflex to the lumbro-sacral region and the spinal cord. This in turn activates the pontine micturition center that communicates with the higher brain center consisting namely of the medial prefrontal cortex (mPFC), anterior cingulate cortex (ACC), and the insular cortex (insular Cx) that promote the guarding reflex for continence. In response, the PAG sends information back to the spinal cord, lumbro-sacral region, bladder to contract the detrusor muscle in the bladder and relax the sphincter muscle in the urethra for voiding. The conscious sensation of bladder pressure and voiding requires the decision of the frontal cortex and the motor learning coordinated through the cerebellum and brainstem. It is the link between the higher centers in the brain with the lower center in the bladder-lumbro-sacral region that constitute the

http://dx.doi.org/10.5772/54947

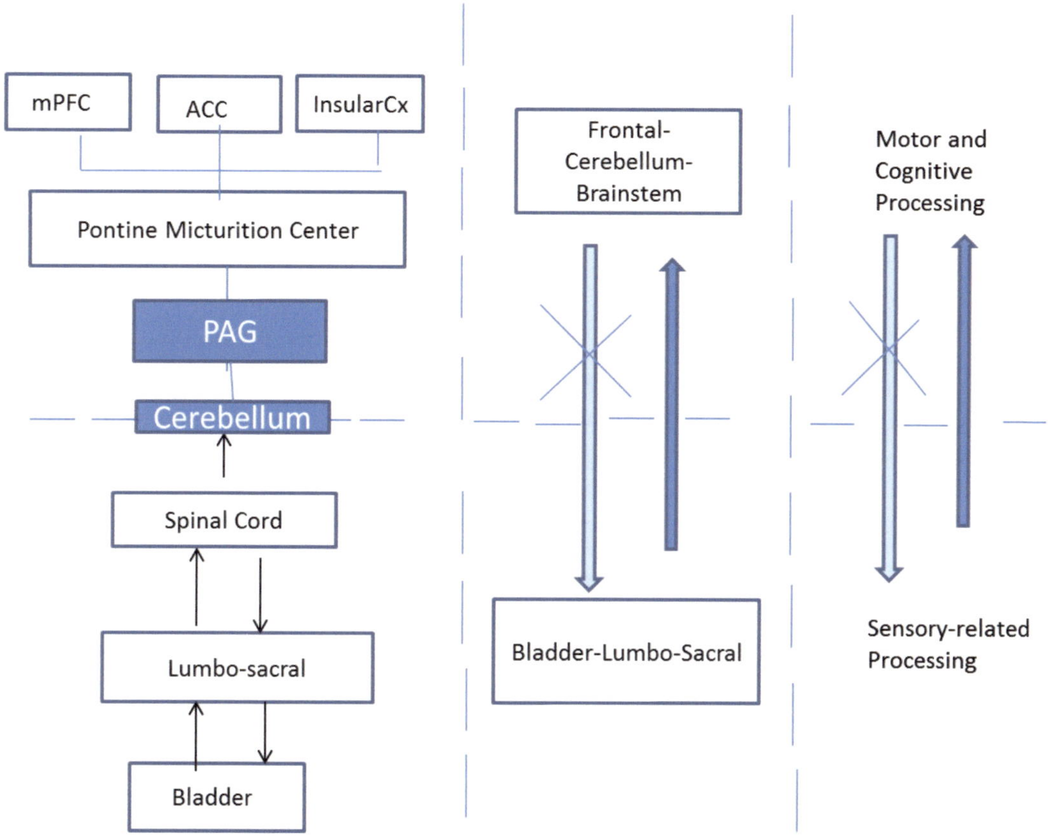

Figure 2. Neural circuits involving the brain that forms voluntary control of micturition.

conscious control necessary for toilet training. In behavior terms, the motor and cognitive learning sub-served by the higher brain centers talks to the lower sensory-related processing in the bladder region to allow the individual to decide whether or not to void. When there is neurological damage as in spinal cord injury or hypothetically in neurodevelopmental disorders such as autism, there is a lack of information flow between the higher (frontal-cerebellum-brainstem) and lower centers (bladder-lumbo-sacral); it may be likely that individuals resistant to toilet training belong to this cohort.

6. Sensory processing in autism

Sensory functions play an important role in young children's daily performance. As defined in the *Occupational Therapy Practice Framework: Domain and Process* (American Occupational Therapy Association [AOTA], 2008), the client factors of sensory functions include touch, smell, taste, hearing, seeing, temperature, pressure, proprioceptive processing and vestibular function. Professionals and parents are recognizing that when young children have poor sensory processing functions, it can affect their sensorimotor, cognitive, and social development (Sears, 1994). ASD are associated with differences in sensory functions and affective response to sensory stimuli. Studies have shown that 70%-100% children with ASD exhibit sensory processing difficulties (Adamson, O'Hare & Graham, 2006; Ben-Sasson et al., 2007; Tomchek & Dunn, 2007). Three sensory constructs (under-responsiveness, over-responsiveness, and sensation seeking) generally characterize the ASD population which are different than typically developing children (Ben-Sasson et al., 2009; Tomchek & Dunn, 2007). These sensory processing difficulties may further impact family routines and daily occupations, such as what a family chooses to do or not to do during *family time* and how the family prepares daily tasks (Bagby, Dickie & Baranek, 2012).

The cerebellum is strongly implicated in the control of continence by relaying sensory impulses from the sacral spinal regions to the higher centers. A dominant role of the cerebellum is the processing of sensory information through the cerebello-thalamo-cortical tract. Aberration in neuronal circuits of the cerebellum have been proposed (Yip et al., 2007, 2008, 2009) and could be an underlying factor for altered sensory sensitivity observed in individuals with ASD.

In addition, there is an association between the modes of sensory processing with temperament (Brock et al., 2012). The ASD population displays temperament scores distinct from typically developing children on most dimensions of temperament (such as activity, rhythmicity, adaptability, approach, distractibility, intensity, persistence, and threshold). Sensory under-responsiveness was associated with slowness to adapt, low reactivity, and low distractibility. Conversely, sensory over-responsiveness often accompanies hyperactivity, attention deficit and high distractability. A combination of increased sensory processing difficulties, especially in the areas of taste & smell sensitivity and movement-related sensory behavior, was associated with greater challenge in self-care skills, adaptive behaviors and emotional regulations (Lane, Young, Baker & Angley, 2010).

Successful integration of signals from various sensory systems is essential for daily functioning. Sensory processing powerfully influences the channel of information transfer necessary for the biology of continence. Sensory processing further contributes critical sensory integration and necessary motor skills for everyday activities and elaborates the temperament of the child, such as to dictate the affective and attentional capability, without which the behaviors associated with toileting could not be accomplished.

7. Analysis of sensory constructs related to toileting in children with ASD

- Under-responsiveness: not aware of wet or stool diapers, not notice when hands are dirty or messy
- Over-responsiveness or sensory sensitivity: overwhelmed by tactile inputs from underpants (or touch feelings of underwear edges), smell of urine, sight of fluorescent lights in bathroom, cold feel of toilet seat, loud noise associated with flushing, etc.

http://dx.doi.org/10.5772/54947

- Sensation seeking: playing water, urine or stool in the toilet, playing water at the sink, turning on and off the lights in the bathroom, flushing toilet repeatedly

All of these sensory processing difficulties may make toilet training a challenge.

8. Analysis of motor and cognitive processing related to toileting in children with ASD

The cerebellum is a critical brain area for the relaying of sensory information to the cerebral cortex. Sensory information conveyed through motor movement is interpreted in the cerebral cortex, which eventually become stored as motor learning. The learning of motor skill is essential to daily self-help routine. In the cerebellum, the olivocerebellar pathway is a key pathway contributing to learning of motor skills. When the olivocerebellar circuit is deficient as reported in the autism literature (Blatt et al., 2010), it is expected that motor learning will correspondingly be impacted. The role of motor learning in neurodevelopmental disorders and its noticeable implications for the understanding and management of these patients has recently gained more attention (Manto & Jissendi, 2012).

It is common for children with ASD to exhibit deficits in sensory-motor processing such as apraxia. For example, complex motor planning and multiple sensory-motor steps are involved in self-care tasks such as undressing, toileting, cleaning up and hand-washing. The task of toileting requires children to utilize sufficient balance, trunk stability & gross motor skills to get on/off toilet seat and fine more coordination to dress & undress. Dressing or undressing is one specific activity within the self-care area commonly known as activity of daily living (ADL). Dressing independently as a child is important for demonstrating self-reliance (Christiansen & Matuska, 2004). Children with ASD often struggle completing these multiple sensory-motor steps related to toileting.

Independent toileting requires many motor skills that differ by gender, by the facility used, and by the type of clothing worn in different contexts followed by specific cultural norms. A child has to demonstrate his or her abilities to accomplish a series of sensory-motor activities. These activities often require a higher level of executive functioning to form an idea of motor sequences, and to perform a complex loop of motor planning & executions of movement patterns. The following paragraphs provide examples of motor and cognitive skills related to toileting:

- Motor skills related to toileting: moving on and off the toilet; sufficient balancing skills; muscle strength, postural control, and trunk stability to sit safely on the toilet; shifting weight to wipe and to undress then dress after toileting; active range of motion to reach for toilet paper, to wipe the perineal area and to flush the toilet; fine motor skills and bilateral coordination to grasp toilet paper; somatosensory and proprioceptive skills to feel the hold/grasping of the toilet paper; directional movements to clean the perineal area; fine motor coordination to unfasten, remove, and refasten underwear and clothing. For boys who have insufficient gross motor skills and are lacking balancing skills or trunk stability may have a difficulty to stand in front of a urinal or direct urine into the toilet. Belts or additional accessories may also be worn, which would require additional eye-hand coordination and perceptual functions.

- Cognitive functions in toileting: familiarizing the child to the environment of the toilet so that it is perceived as being a friendly place where one uses the toilet for actions associated with voiding and evacuation; establishing a habit of sitting on the toilet for a length of time; succesfully voiding and evacuating in the toilet and being reinforced positively; forming a habit of going to the toilet to void and evacuate; realizing the urge to void and independently going to the toilet; feeling the pressure to the bowel and independently making the trip to the toilet; performing the steps of voiding, evacuation, cleaning up oneself, handwashing and drying the hands.

9. Task analysis of toileting in children with ASD

Toilet hygiene is an important activity of daily living task as it relates to independence in daily functioning, self-identity, and social acceptance. The following paragraphs provide examples of how toileting may impact routines and psychosocial aspects in children with ASD:

- Social Interaction and Imitation: interest in watching others perform certain tasks, then imitating "go potty", in children with ASD less responsive to social rewards such as praise of "good job" for successful toileting might be a challenge.
- Communication: ability to express a need to go to the bathroom. Boswell & Gray (2012) in TEACCH indicate the importance of having a communication system for children with ASD to initiate the toilet sequences, such as signaling that he needs to go to the bathroom; and have a back up plan in place when in a unfamiliar environment. Such plan may include an evaluation of the environmental contexts and incorporating "teachable moments" to teach the child how to use systematic communication tools, such as objects, pictures, or words to communicate in other settings.
- Tasks in bathroom include: obtaining and using supplies, clothing management – undressing and dressing, maintaining toileting position, transferring to & from toileting position, cleaning body & maintaining posture & balance during cleaning, caring for menstrual needs (Christiansen & Matuska, 2004).
- Teaching and practicing toilet hygiene: children with ASD who also exhibit tactile sensitivity may avoid hand-over-hand assistance to turn the water on and off at the sink during handwashing (Kern, Wakeford & Aldridge, 2007; Dunn, 1997). With compromised motor planning and eye-hand coordination, the movement of turning on and off of light and water, or holding on the handle to flush toilet would be daunting for the children with ASD.
- Change in routines associated with toileting: taking away diapers, wearing underpants.
- Caregiving: there are certain caregiving stresses associated with toileting (Dalrymple & Ruble, 1992). When a child is toilet trained, financial and time benefits arise, such as money

savings of diapers and wipes; available time from assisting toileting for caregivers to engage in other daily tasks (Lee, 2010).

The next paragraph implements, via case studies, the aforementioned task analysis involved in toilet training, including the motor skills and cgnitive domain. Case studies of two children who are siblings illustrate the success of toilet training in one child and a lack of independent toileting in another child; therefore, provide insight into factors that could hinder success.

10. Case studies utilizing the training protocol

The procedure is adapted from Azrin and Foxx (1971) and was carried out in the children's home during the evening 4 hours before bedtime. The participants' parents were interviewed prior to baseline collection and the habit journal was constructed based on the interview. In addition, the researcher compiled a journal recording the number and time of voids after mealtime and at each hour noting the baseline voiding pattern. The data was collected across two children, who are siblings, given the pseudo-name of Jim (10 year old male with a diagnosis of autism) and Bob (5 year old male also with autism) for anonymity. A schedule of reinforcers were noted and used for the protocol (see Table 1). Baseline data includes the frequency of voids per trip to the toilet. Both children were able to void, on occasion, upon verbal prompting but self-initiation was lacking. After baseline was taken, the toilet training program commenced. Unfortunately the training was not as intensive as would have been the ideal scenario. The training program was limited logistically by the children's schedule and school commitment, resulting in the time allocated for the toilet training programme to be 3 hours per evening for twice a week. After 4 weeks at this level of contact time, voiding was 100% for both children (Figures 1 and 2). Follow up after the intervention was maintained at 100%. However, self-initiation was only observed in Bob but not in Jim. While both children have autism, Bob's level of communication was superior to Jim in that Bob could independently express approximately 15 words. On the other hand, Jim was non-verbal and demonstrated less cognitive skills and receptive communication.

The protocol consists of first familiarizing the children to the toilet routine. Children were shown videos, pictures, songs of all aspects associated with toileting. After the multi-media exposure, children were encouraged to visit the bathroom, play with flushing the toilet, turn on and off the faucet, wash their hands with soap, dry with a towel and play with the toilet paper. The aim of the first stage is to allow the children to enjoy going to the bathroom. Stage 2 of the protocol is the child entering the bathroom in the presence of reinforcers so that the bathroom is viewed as a friendly place. Children sat on the toilet for 5 minutes with their favourite toy and activity. The time is increased to 10 minutes when children showed enthusiasm for the toilet sitting routine. When successful, implementation of the scheduled sitting begins. The routine of 15 minutes on the toilet followed by 15 minutes of break away from the toilet and bathroom was performed. To provide maximum opprtunity for voiding, children were given drinks every 15 minutes. In scheduled toilet sitting, the children were taught to undress by themselves from the waist down and continuously sit on the toilet without reinforcer. As they void, they were praised and reinforced with their favourite toy. As the percentage of voids increases, the children were given breaks even before the 15 minute-sitting time were up to instill the notion that once voiding has completed, it is time to leave the bathroom. Upon independent void, children were immediately rewarded with fun and interesting activities in another room.

• Stage 1: Introduction to toilet through multi-media education, such as Elmo potty video, toilet training songs, Elmo coloring book, toilet training stories
• Stage 2: 5 minutes on the toilet engaged in favourite activity, such as watching DVD, playing Lego®, or favourite toys
• Stage 3: Extend to 10 minutes on the toilet engaged in favourite activity
• Stage 4: 15 minutes uninterrupted toilet sitting while engaged in favourite activity
• Stage 5: Hydrate with water every 15 minutes followed by routine 15 minutes toilet sitting
• Stage 6: Toilet sitting for 15 minutes duration, without favourite activity, intervened with 15 minutes break. This is the scheduled 15 minutes on toilet and 15 minutes off toilet
• Stage 7: Toilet sitting for 15 minutes noting successful voids and reinforcing with praise and 2 minutes of favourite activity as reinforcer and a break
• Stage 8: When voiding is independent and successful, breaks are introduced into the 15 minutes toilet sitting and 15 minutes break routine extending the time of break

Table 1. Protocol for Voiding

Jim and Bob differed in their outcomes in toilet training. While both children could complete the steps associated with toileting as shown in Figures 3a and 4a, only Bob could initiate the process independently and did not need to wear a diaper following intervention since Bob could initiate independently (see Figure 4b). The other child, Jim, could only void when prompted and failed to achieve the independence (Figure 3b) necessary in order for him to graduate from the diaper. A sensory profile analysis (elaborated in succeeding paragraphs) using the Sensory Profile and the Sensory Processing Measure (SPM) illustrated the difference in sensory processing between Bob and Jim. Bob, not only has better cognitive ability but also has better sensory processing which may contribute to a more intact neural circuit between the higher brain centers with the lower bladder-lumbro-sacral and spinal region. It is likely that an intact circuit predicts better outcomes in toilet training. Nevertheless, further sensory and cognitive training for Jim could help him eventually develop independence.

11. Evaluation Results and Summary Interpretation of Sensory Processing for the Case Studies

11.1. Sensory profile[1]

The Sensory Profile Caregiver Questionnaire (SP; Dunn, 1999) is a widely used pediatric assessment that provides a standard method for professionals and caregivers to measure

1 Dunn, W. (1999).Sensory Profile user's manual.San Antonio, TX:Psychological Corporation.

a)

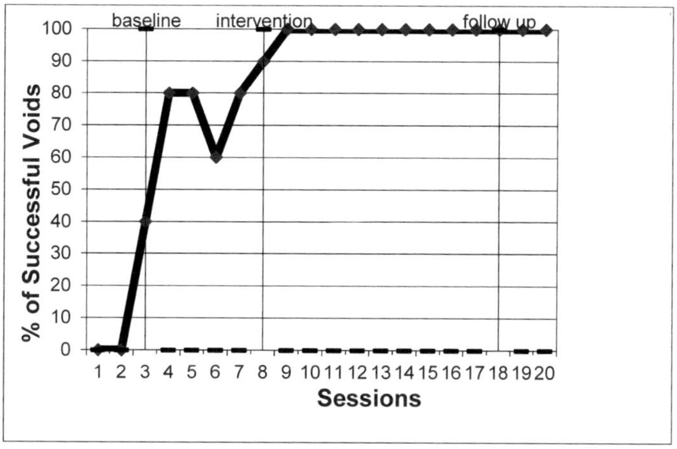

b)

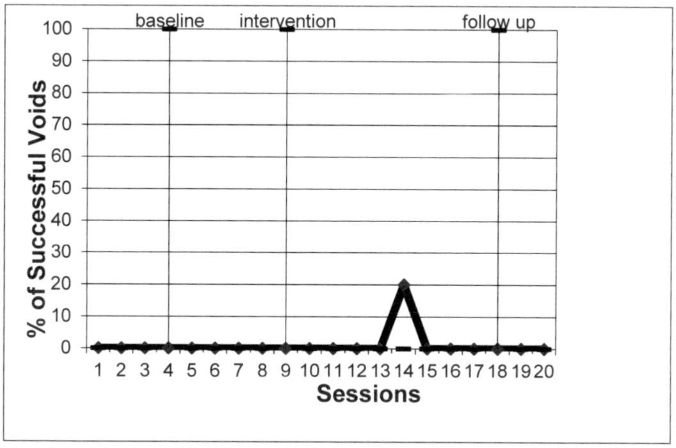

Figure 3. Percentage of successful voids (3a) and percentage successful void upon self-initiation (3b) for Jim.

children's (from 3-10 years of age) responses to sensory events in everyday experiences. Caregivers complete the questionnaire by reporting how frequently their children respond in the way described by each item, using a 5-point Likert scale (nearly never, seldom, occasionally, frequently, almost always). There are 125 items included in the profile[2], which are: (1) sensory processing, (2) modulation, and (3) behavioral & emotional responses. The sensory

2 Cut off scores were determined by preparing a cumulative frequency distribution with the national research sample of children without disabilities (n = 1,037) (Dunn & Westman, 1997) and computing the raw scored cut scores for -1 SD and -2 SD. This was done for the SSP total score and for each section. Refer to the Summary section for representation of the raw scores for 1 SD and 2 SD below the mean.

a)

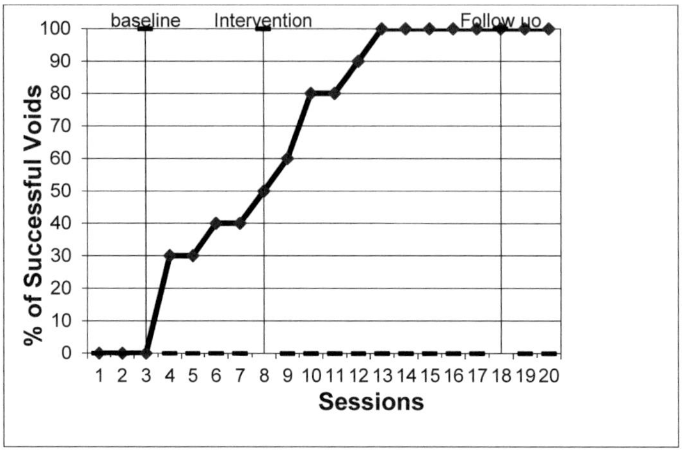

b)

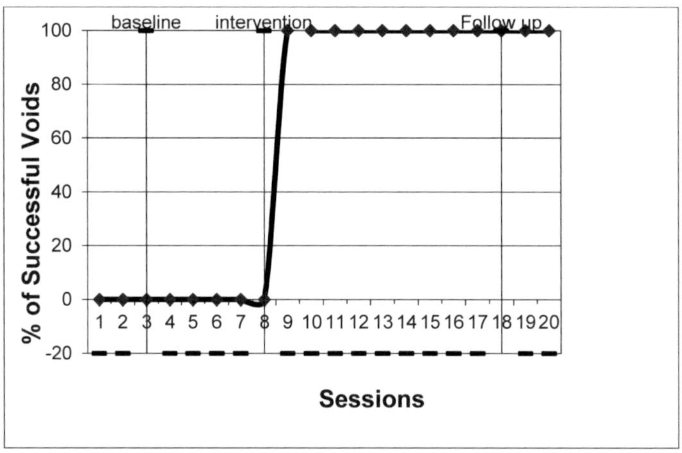

Figure 4. Percentage of successful voids (4a) and percentage successful void upon self-initiation (4b) for Bob.

processing section includes six sensory systems (e.g., auditory, visual, vestibular, tactile, oral). The modulation section contains five item categories that measure children's ability to regulate sensory inputs in order to produce an appropriate response to the context (Dunn, 1997). The summary scores are then classified as *Typical Performance*[3], *Probable Difference*[4], or *Definite Difference*[5] in each sensory processing system. The Sensory Profile was given to Jim and Bob's

3 Typical Performance is scores at or above the -1 SD point. This classification system is a statistical proportion usually represents about 84% of the norm. Section raw score totals that fall within this range indicate typical sensory processing abilities. This range indicates that the child performed no different than the 84% of the research sample of childrenwithout disabilities (n = 1,037).

http://dx.doi.org/10.5772/54947

caregiver to reveal the possible contribution of his sensory processing patterns to daily performance and/or functional challenges. The following paragraphs summarize Jim and Bob's performance from the *Sensory Profile*.

11.2. Jim's section & factor summaries

Jim has Probable Difference scores in the following areas:

- Modulation Difficulty in *Visual Processing*: occasionally prefers to be in the dark; has difficulty putting puzzles together (as compared to same age children); has a hard time finding objects in competing backgrounds (for example, shoes in a messy room, favorite toy in the junk drawer); occasionally looks carefully or intensely at objects/people (such as stares); has trouble staying between the lines when coloring or when writing.
- Modulation Difficulty in *Touch Processing*: doesn't seem to notice when face or hands are messy (low registration); has difficulty standing in line or close to other people (tactile sensitivity); expresses discomfort at dental work or tooth brushing (for example, cries or fights); avoids getting messy (such as in paste, sand, finger paint, glue, tape).
- *Multisensory Processing*: has difficulty paying attention; walks on toes.
- Modulation Related to *Body Position and Movement*: fears falling or heights; seems accident-prone; hesitates going up or down curbs or steps (for example, in cautious, stops before moving).
- Modulation of *Movement* Affecting *Activity Level*: Low Endurance/Tone; spends most of the day in sedentary play (for example, does quiet things); becomes overly excited during movement activity.
- Modulation of *Visual Input* Affecting *Emotional Responses* & *Activity Level*: avoids eye contact; watches everyone when they move around the room; doesn't notice when people come into the room.
- Emotional/Social Responses: has trouble "growing up" (for example, reacts immaturely to situations); has definite fears; has difficulty making friends (for example, does not interact or participate in group play); doesn't have a sense of humor.
- Sensory Seeking: enjoys strange noises/seeks to make noise for noise's sake; seeks out all kinds of movement activities.

Jim has Definite Difference scores in the following areas:

---

4 Probable Difference is scores at or above the -2 SD point below the mean, but lower than -1 SD point below the mean. This classification system is a statistical proportion usually represents about 14% of the norm. Section raw score totals that fall within this range indicate questionable areas of sensory processing abilities. This range indicates that the child's performance was between the 2nd and 16th percentile, representing 14% of the sample population.

5 Definite Difference is scores below the -2 SD point below the mean. This classification system is a statistical proportion usually represents about 2% of the sample population. Section raw score totals that fall within this range indicate sensory processing difficulties. This range indicates that the child is performing like a child in the lowest 2% of the sample, when compared to the research sample of 1,037 children without disabilities.

- *Auditory Hypersensitivity*: responses negatively to unexpected or loud noises (for example, cries or hides at noise from vacuum cleaner, dog barking, hair dryer); has trouble completing tasks when the radio is on.
- *Seeking Vestibular Inputs*: seeks all kinds of movement and this interferes with daily routines (for example, cannot sit still, fidgets); occasionally rocks in desk/chair/on floor.
- *Oral Motor Hypersensitivity* and *Dyspraxia*: avoids certain tastes or food smells that are typically part of children's diets; will only eat certain tastes (such as rice, fish & chicken); picky eater, especially regarding food textures.
- Modulation Related to *Endurance/Tone*: cannot lift heavy objects; shows poor endurance & tires easily.
- Modulation of Sensory Inputs Affecting *Emotional/Social Responses and Daily Performance*: such as *Poor Registration* and *Oral Sensory Sensitivity*; does perceive body language or facial expressions (for example, unable to interpret).
- Behavioral Outcomes due to Challenges in Sensory Processing: such as Emotionally Reactive and Inattention/Distractibility

11.3. Bob's section & factor summaries

Bob has Probable Difference scores in the following areas:

- Sensory Seeking in *Visual Processing*: seeks all kinds of movement and this interferes with daily routines (for example, can't sit still, fidgets, being whirled by adult, merry-go-rounds, playground equipment, moving toys, etc.).
- *Oral Motor Hypersensitivity* and *Dyspraxia*: gags easily with food textures or food utensils in mouth; avoids certain tastes or food smells that are typically part of children's diets; will only eat certain tastes (such as rice, fish, chicken & noodle); picky eater, especially regarding food textures.
- Modulation Related to *Body Position and Movement*: seems accident-prone; hesitates going up or down curbs or steps (for example, in cautious, stops before moving); fears falling or heights; takes movement or climbing risks during play that compromise personal safety.
- Sensory Seeking: enjoys strange noises/seeks to make noise for noise's sake; seeks out all kinds of movement activities; becomes overly excited during movement activity; "on the go".

Bob has Definite Difference scores in the following areas:

- Modulation of *Auditory Processing*: holds hands over ears to protect ears from sound; responses negatively to unexpected or loud noises (for example, cries or hides at noise from vacuum cleaner, dog barking, hair dryer); has trouble completing tasks when the radio is on; appears to not hear what you say (for example, does not "tune-in" to what you say, appears to ignore you).

- *Multisensory Processing*: has difficulty paying attention; looks away from tasks to notice all actions in the room.
- Modulation of *Movement* Affecting *Activity Level*: spends most of the day in sedentary play (for example, does quiet things); becomes overly excited during movement activity; frequently "on the go".
- Modulation of *Sensory Inputs* Affecting *Emotional/Social Responses* and *Daily Performance*: express discomfort at dental work or tooth brushing; don't seem to notice when face or hands are messy; has difficulty making friends (for example, does not interact or participate in group play); talks self through tasks; has trouble staying between the lines when coloring or when writing; has difficulty tolerating changes in routines.
- *Inattention/Distractibility*: contributed by poor modulation of *Auditory Processing*
- *Low Registration*: high pain tolerance; lack of temperature awareness; don't seem to notice when face or hands are messy.

12. Sensory profile results

In comparison of Jim & Bob's Sensory Profile results, Jim demonstrated multifaceted sensory processing and modulation difficulties in several areas, including: *Low Endurance/Tone, Oral Sensory Sensitivity, Fine Motor/Perceptual Skills, Visual Processing, Vestibular Processing,* and *Modulation of Visual Input Affecting Emotional Responses and Activity Level.* Despite showing a more organized sensory processing profile, Bob displays problems in *Modulation of Movement Affecting Activity Level* and *Emotional/Social Responses*. Through sensory motor activities, he demonstrates ideational dyspraxia and problems with motor planning, which are validated through his Sensory Profile by engaging in sedentary play. In comparison to Jim, who displays complex multi-sensory processing deficits, including somatodyspraxia, tactile, vestibular & proprioceptive among other sensorimotor deficits. This may help explaining Jim's higher level of difficulties with activities of daily living (ADLs), including toileting.

13. Sensory Processing Measure (SPM)[6]

The Sensory Processing Measure (SPM) is an integrated system of rating scales that is used to assess children's response to sensory stimuli in various environments. The SPM consists of three forms: the Home Form, the Main Classroom Form, and the School Environments Form. The SPM is based on sensory integration theory (Ayres, 1972[7], 1979[8], 2005[9]), which proposes that the integration and regulation of everyday sensory experiences is critical neurobehavioral

---

6 Parham, L. D., Ecker, C., Miller Kuhaneck, H., Henry, D. A., & Glennon, T. J. (2007). Sensory Processing Measure (SPM): Manual. Los Angeles, CA: Western Psychological Services.

7Ayres, A. J. (1972). Sensory integration and learning disorders. Los Angeles, CA: Western Psychological Services.

Factor Summary		
Category	**Raw Score - Jim**	**Raw Score - Bob**
Sensory Seeking	61/85*	58/85*
Emotional Reactive	46/80**	34/80**
Low Endurance/Tone	37/45*	44/45
Oral Sensory Sensitivity	24/45**	32/45*
Inattention/Distractibility	16/35**	15/35**
Poor Registration	24/40**	27/40**
Sensory Sensitivity	15/20*	16/20
Sedentary	10/20*	8/20**
Fine Motor/Perceptual	8/15*	12/15
Section Summary		
Sensory Processing		
Category	**Raw Score - Jim**	**Raw Score - Bob**
Auditory Processing	19/40**	16/40**
Visual Processing	30/45*	35/45
Vestibular Processing	43/55**	47/55*
Touch Processing	73/90	75/90
Multisensory Processing	24/35*	22/35**
Oral Sensory Processing	34/60**	41/60*
Modulation		
Category	Raw Score - Jim	Raw Score - Bob
Sensory Processing Related to Endurance/Tone	32/45**	44/45
Modulation Related to Body Positioning Movement	40/50*	40/50*
Modulation of Movement Affecting Activity Level	19/35*	14/35**
Modulation of Sensory Input Affecting Emotional Responses	10/20**	11/20**
Modulation of Visual Input Affecting Emotional Responses and Activity Level	13/20*	16/20
Behavior and Emotional Responses		
Category	Raw Score - Jim	Raw Score - Bob
Emotional/Social Responses	55/85*	43/85**
Behavioral Outcomes of Sensory Processing	15/30**	15/30**
Items Indicating Thresholds for Response	13/15	13/15

No indication: Typical Performance

* Indicates *Probable Difference*

** Indicates *Definite Difference*

Table 2. Comparison of Jim & Bob's Sensory Profile results (comparing to children aged 3-10 years old):

8Ayres, A. J. (1979). Sensory integration and the child. Los Angeles, CA: Western Psychological Services.

process that affects development and function. Reponses of "never", "occasionally", "frequently", or "always" are given in answer to questions regarding areas of *Social Participation, Vision, Hearing, Touch, Taste & Smell, Body Awareness, Balance & Motion*, and *Planning & Ideas*. Results are converted to T scores, and compared to responses of typically developing children aged 3 to 12 years old. Scores are interpreted within ranges described as "typical" (within normal limits), "some problems" (*T score of 60-69), and "definite dysfunction" (**T score of 70-80). In addition to the Sensory Profile, Jim and Bob's caregiver also completed Sensory Processing Measure (SPM) – Home Form[10]. This assessment is similar to the Sensory Profile, with additional information on *Social Participation* and *Praxis* (the ability to plan and organize movement).

Results of the SPM (Table 3) indicate Jim is experiencing definite dysfunction in the areas of *Hearing, Motor Planning,* and *Social Participation*. Some sensory problems are also indicated, including: *Vision, Touch, Body Awareness,* and *Balance & Motion*. Bob's SPM results show fewer deficits in multi-sensory processing, with definite dysfunction in *Motor Planning & Ideas*. Some sensory problems are indicated, including: *Vision, Hearing, Touch, Motor Planning,* and *Social Participation*. In comparison of Jim & Bob's Sensory Processing Measure results, Jim demonstrates more sensory deficits in multiple categories, especially in the areas of *Social Participation, Hearing,* and *Body Awareness*.

	Raw Score		**T-Score**		**Percentile**	
Area of Assessment	**Jim**	**Bob**	**Jim**	**Bob**	**Jim**	**Bob**
Social Participation	38	25	80**	65*	"/>99	93
Vision	23	21	69*	68*	97	96
Hearing	20	18	71**	69*	98	97
Touch	17	21	61*	66*	86	95
Body Awareness	17	14	61*	57	86	76
Balance & Motion	19	20	64*	65*	92	93
Planning/Ideas	35	25	80**	70**	"/>99	97.5
Total	104	103	67*	67*	95.5	95.5

T-Score of 40-59 indicates *"typical performance"*

*T-Score of 60-69 indicates *"some problems"*

**T-Score of 70-80 indicates *"definite dysfunction"*

Table 3. Comparison of Jim & Bob's Sensory Processing Measure (SPM): Home Report Results

9Ayres, A. J. (2005). Sensory integration and the child, 25th anniversary edition. Los Angeles, CA: Western Psychological Services.

10Parham, L. D., & Ecker, C. (2007). Sensory Processing Measure (SPM) Home Form. Los Angeles, CA: Western Psychological Services.

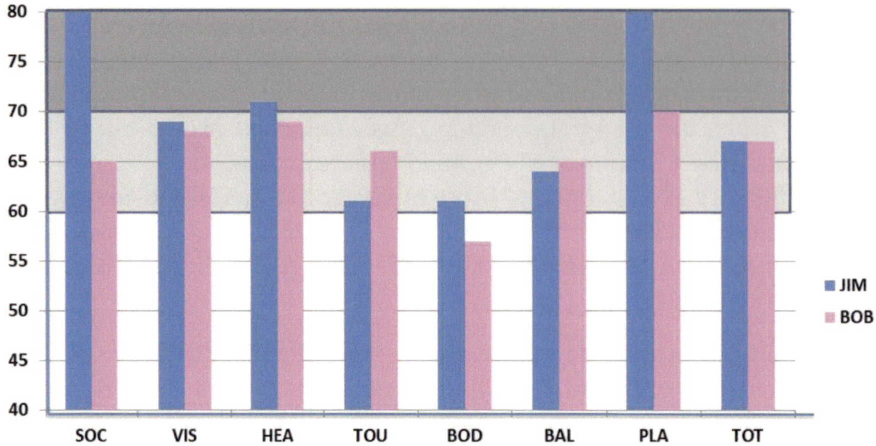

Figure 5. A summary T-Scores of Sensory Processing Measure (SPM) for Jim and Bob. Sensory processing areas are represented on the x-axis and T-Scores are indicated on the y-axis. The darkest shaded zone for the T-Scores between 70 and 80 represent *definite dysfunction*. Correspondingly, the lighter shaded zone between T-Scores of 60 and 70 indicate *some problems*. Note that although the total T-Scores for Jim and Bob is the same and fell under "some problems", Jim has a more complex sensory processing profile and somatosensory processing deficits as indicated on the above-mentioned Sensory Profile results.

14. Summary interpretation of Jim and Bob's sensory processing

When a child has difficulty in a particular sensory system, it means that this form of sensory input would be confusing or upsetting to the child. As a result, the child may avoid situations that overload him with specific types of sensory stimuli or acting inappropriately when such stimuli are unavoidable. Problems in multisensory processing could cause a child to become easily agitated or distracted around busy environments & while experiencing new sensations. At times, the child may show emotional distresses and seem disorganized in their movement patterns or social interactions. In any case, difficulty with sensory inputs can interfere with the child's ability to modulate emotions, behavior, and/or activity level appropriately in response to sensory stimuli in contexts and to complete important activities successfully.

The Sensory Profile & the Sensory Processing Measure scores reveal Jim and Bob's sensory processing and modulation difficulties in multiple areas. Jim's sensory modulation difficulties are more profound and complex than Bob (Table 2, Table 3 and Figure 5). For example, Jim demonstrates several sensory-based motor deficits and somatodyspraxia, including tactile defensiveness, poor endurance, low muscle tone, and lack of body awareness. His proximal joint instability, poor posture control & lack of trunk stability add to his uncoordinated movements & fatigue during gross motor tasks. In order to make sense of his body positions in space, Jim may exhibit excessive movements, grasp objects too tightly, or push too hard when engaging with everyday objects. In comparison, Bob's ideational dyspraxia may contribute to his needs of repeating movement patterns to make sense of movement patterns.

http://dx.doi.org/10.5772/54947

These sensory processing deficits would affect their attention to tasks, social participation and functional performance in daily routines, such as toileting.

Future directions

Much of the research studies on toilet training in autism has centered on operant conditioning, especially training protocols. Toilet training develops as a result of culture and is a condition of civilized living, but there may be reasons for examining an even earlier process. This process speaks of the inherent biological drive which communicates between lower and higher centers and is shaped by learning. In the population of ASD who may be cognitively impacted to a severe degree, independence in toileting is inhibited by the lack of sensory-neural integration. This chapter attempts to address the issue of sensory processing challenges in toilet training for the intervention of the resistant group. Toileting is an essential daily living skill which typically developing individuals often take for granted. It is in the lack of its skill acquisition that profoundly impacts the children with ASD, their families, caregivers and society as a whole.

Abbreviations

SOC = Social Participation

VIS = Vision

HEA = Hearing

TOU = Touch

BOD = Body Awareness

BAL = Balance and Motion

PLA = Motor Planning and Ideas

TOT = Total T-Scores

Author details

Jane Yip[1,3], Betsy Powers[2] and Fengyi Kuo[1]

1 Autism Parent Care, USA

2 Indiana University School of Health and Rehabilitation Sciences, USA

3 Purdue University, USA

References

[1] Adamson, A., O'Hare, A., & Graham, C. (2006). Impairments in sensory modulation in children with autistic spectrum disorder. *British Journal of Occupational Therapy, 68(*8), 357-364.

[2] American Occupational Therapy Association. (2008). Occupational therapy practice framework: Domain and process (2nd ed.). *American Journal of Occupational Therapy, 62,* 625–683. doi: 10.5014/ajot.62.6.625

[3] Ayres, A. J. (1972). *Sensory integration and learning disorders.* Los Angeles, CA: Western Psychological Services.

[4] Ayres, A. J. (1979). *Sensory integration and the child.* Los Angeles, CA: Western Psychological Services.

[5] Ayres, A. J. (2005). *Sensory integration and the child, 25th anniversary edition.* Los Angeles, CA: Western Psychological Services.

[6] Azrin, N. H., & Foxx, R. M. (1971). A rapid method of toilet training in the institutionalized retarded. *Journal of Applied Behavior Analysis, 4,* 89-99.

[7] Bagby, M. S., Dickie, V. A., & Baranek, G. T. (2012). How sensory experiences of children with and without autism affect family occupations. *American Journal of Occupational Therapy, 66,* 78–86. doi: 10.5014/ajot.2012.000604

[8] Ben-Sasson, A., Cermak, S. A., Orsmond, G. I., Tager-Flusberg, H., Carter, A. S., Kadlec, M. B., & Dunn, W. (2007). Extreme sensory modulation behaviors in toddlers with autism spectrum disorders. *American Journal of Occupational Therapy, 61,* 584–592.

[9] Ben-Sasson, A., Hen, L., Fluss, R., Cermak, S., Engel-Yeger, B., & Gal, E. (2009). A meta-analysis of sensory modulation symptoms in individuals with autism spectrum disorders. *Journal of Autism and Developmental Disorders, 39(*1), 1-11.

[10] Blatt, G. J., Soghomonian, J. J., & Yip, J. (2010). Glutamic Acid Decarboxylase (GAD) as a Biomarker of GABAergic Activity in Autism: Impact on Cerebellar Circuitry and Function. In G. J. Blatt (Ed.), *The Neurochemical Basis of Autism* (1st ed.). New York; Springer.

[11] Boswell, S., & Gray, D. (2012). *Applying structured teaching principles to toilet training.* Retrieved May 4, 2012 from http://teacch.com/educational-approaches

[12] Brock, M. E., Freuler, A., Baranek, G. T., Watson, L. R., Poe, M. D., & Sabatino, A. (2012). Temperament and sensory features of children with autism. *Journal of Developmental Disorders.* Feb 25 (Epub ahead of print) doi: 10.1007/s10803-012-1472-5

[13] Cermak, S. A., Orsmond, G. I., Tager-Flusberg, H., Carter, A. S., Kadlec, M. B., & Dunn, W. (2007). Extreme sensory modulation behaviors in toddlers with autism spectrum disorders. *American Journal of Occupational Therapy, 61*, 584–592.

[14] Chapparo, C. J., & Hooper, E. (2005). Self-care at school: Perceptions of 6-year-old children. *American Journal of Occupational Therapy, 59*, 67–77.

[15] Christiansen, C. H., & Baum, C. M. (2005). The complexity of human occupation. In C. H. Christiansen, C. M. Baum, & J. Bass-Haugen (Eds.), *Occupational therapy: Performance, participation, and well-being* (3rd ed.) (p. 9). Thorofare, NJ: SLACK Inc.

[16] Christiansen, C. H., & Matuska, K. M. (2004). *Ways of living: Self-care strategies for special needs* (3rd ed.). Bethesda, MD: American Occupational Therapy Association.

[17] Cicero, F. R., & Pfadt, A. (2002). Investigation of a reinforcement-based toilet training procedure for children with autism. *Research in Developmental Disabilities, 23*, 319-331.

[18] Dalrymple, N. J., & Ruble, L. A. (1992). Toilet training and behaviors of people with autism: Parent views. *Journal of Autism and Developmental Disorders, 22*, 265-275.

[19] Dunn, W. (1997). The impact of sensory processing abilities on the daily lives of young children and their families: A conceptual model. *Infants & Young Children, 9(*4), 23-35.

[20] Fowler, C. J., Griffiths, D. & de Groat, W. C. (2008). The neural control of micturition. *Nat Rev Neurosci, 9(*6), 453-466.

[21] Furman, A. (2001). Young children with autism spectrum disorder. *Early Childhood Connections, 7(*2), 43-49.

[22] Gourcerol, G., Vitton, V., Leroi, A. M., Michot, F., Abysique, A., & Bouvier, M. (2011). How sacral nerve stimulation works in patients with faecal incontinence. *Colorectal Dis. 13(*8), e203-11.

[23] Gray, D. E. (1994). Coping with autism: Stresses and strategies. *Sociology of Health & Illness, 16(*3), 275-302.

[24] Griffiths, D. J., Tadic, S. D., Schaefer, W., & Resnick, N. M. (2009). Cerebral control of the lower urinary tract: How age-related changes might predispose to urge incontinence. *Neuroimage, 47(*3), 981-6.

[25] Kellegrew, D. H. (1998). Creating opportunities for occupation: An intervention to promote the self-care independence of young children with special needs. *American Journal of Occupational Therapy, 52*, 457-465.

[26] Kellegrew, D. H. (2000). Constructing daily routines: A qualitative examination of mothers with young children disabilities. *American Journal of Occupational Therapy, 54*, 252-259.

[27] Kern, P., Wakeford, L., & Aldridge, D. (2007). Improving the performance of a young child with autism during self-care tasks using embedded song interventions: A case study. *Music Therapy Perspectives, 25*(1), 43-51.

[28] Kroeger, K., & Sorensen, R. (2010). A parent training model for toilet training children with autism. *Journal of Intellectual Disability Research, 54*, 556–567. doi: 10.1111/j. 1365-2788.2010.01286.x

[29] Lane, A. E., Young, R. L., Baker, A. E. Z., & Angley, M. T. (2010). Sensory processing subtypes in autism: Association with adaptive behavior. *Journal of Autism Developmental Disorder, 40*, 112–122.

[30] LeBlanc, L. A., Carr, J. E., Crossett, S. E., Bennett, C. M., & Detweiler, D. D. (2005). Intensive outpatient behavioral treatment of primary urinary incontinence of children with autism. *Focus on Autism and Other Developmental Disabilities, 20*(2), 98-105.

[31] Lee, E. B. (2010). *Toilet training and autism spectrum disorders.* Nashville, TN: Vanderbilt University Medical Center.

[32] Manto, M. U., & Jissendi, P. (2012). Cerebellum: links between development, developmental disorders and motor learning. *Front Neuroanat. 6*, 1. Epub Jan. 23

[33] Parham, L. D., & Ecker, C. (2007). *Sensory Processing Measure (SPM) Home Form.* Los Angeles, CA: Western Psychological Services.

[34] Parham, L. D., Ecker, C., Miller Kuhaneck, H., Henry, D. A., & Glennon, T. J. (2007). *Sensory Processing Measure (SPM): Manual.* Los Angeles, CA: Western Psychological Services.

[35] Porterfield, S. L. (2002). Work choices of mothers in families with children with disabilities. *Journal of Marriage and Family, 64*, 972–981.

[36] Rockville, (2006). The effectiveness of different methods of toilet training for bowel and bladder control in agency for healthcare research & quality (US). *Evidence Reports/Technology Assessments, No. 147;* Bookshelf ID: NBK38232.

[37] Rogers, S. J., & D'Eugenio, D. B. (1991). *Developmental Programming for Infants and Young Children: Vol. 1. Assessment and Application.* Ann Arbor, MI; The University of Michigan Press.

[38] Rumfelt-Wright, C. S. (2001). Keeping it clean. *All Together Now! (ATN), 7*(2), 5-7.

[39] Sasa, M., & Yoshimura, N. (1994). Locus coeruleus noradrenergic neurons as a micturition center. *Microsc Res Tech, 29*(3), 226-30.

[40] Sears, C. (1994). Recognizing and coping with tactile defensiveness in young children. *Infant & Young Children, 6*(4), 46-53.

[41] Shelov, S., Altmann, T. R. (Eds.) (2009). *Caring for Your Baby and Young Child: Birth to Age 5* (5th ed.). New York, NY: Bantam Books, American Academy of Pediatrics.

[42] Tomchek, S. D., & Dunn, W. (2007). Sensory processing in children with and without autism: A comparative study using the Short Sensory Profile. *American Journal of Occupational Therapy, 61*, 190–200.

[43] Wheeler, M. (2007). *Toilet training for individuals with autism or other developmental issues (2nd ed.).* Arlington, TX: Future Horizons.

[44] Yip, J., Soghomonian, J. J., & Blatt, G. J. (2007). Decreased GAD67 mRNA levels in cerebellar Purkinje cells in autism: Pathophysiological implications. *Acta Neuropathol, 113(*5), 559-68.

[45] Yip, J., Soghomonian, J. J., & Blatt, G. J. (2008). Increased GAD67 mRNA levels in cerebellar interneurons in autism: Implications to Purkinje cell dysfunction. *J Neurosci Research, 86(*3), 525-530.

[46] Yip, J., Soghomonian, J. J., & Blatt, G. J. (2009). Decreased GAD65 mRNA levels in select subpopulations in the cerebellar dentate nuclei in autism: An *in situ* hybridization study. *Autism Research, 2(*1), 50-9.

Chapter 6

The Recovery Orientation of a Farm Community for Severe Autism — Data from the DREEM-IT (Developing Recovery Enhancing Environment Measures — Italian Version)

Marianna Boso, Enzo Emanuele, Elizabeth Barron,
Noemi Piaggi, Giulia Scanferla, Matteo Rocchetti,
Umberto Provenzani, Davide Broglia, Paolo Orsi,
Roberto Colombo, Sara Pesenti, Marta De Giuli,
Elena Croci, Stefania Ucelli, Francesco Barale,
Jenny Secker and Pierluigi Politi

Additional information is available at the end of the chapter

http://dx.doi.org/10.5772/55007

1. Introduction

Recent years have witnessed an increasing interest in the concept of 'recovery' in the field of mental health and psychiatry. Anthony (1993) described personal recovery as occurring in the presence of ongoing symptoms but involving 'a way of living a satisfying, hopeful and contributing life even with limitations caused by illness' (Anthony, 1993). Recovery from mental illnesses has been conceptualized to involve not only remission of symptoms and achievement of psychosocial milestones but also subjective changes in how persons appraise their lives and the extent to which they experience themselves as meaningful agents in the world (Jacobson & Greenley, 2001). Diverse forms of recovery are possible. In people with optimal outcome, recovery may produce important remission and changes, including the exit from mental health services for a long time period or, sometimes, permanently (Emsley et al, 2011). For other patients, it may mean continuing to receive medical, personal or social support, enabling people to get on with their lives (Emsley et al, 2011). However, in all conditions, the role played by the service in promoting, maintaining and restoring an adequate level of recovery for each patients is pivotal.

Autism is a neurodevelopmental disorder characterized by qualitative impairments in social interaction and communication skill, along with a restricted, repetitive, and stereotyped pattern of behavior and interests (APA, 2000). The diagnosis is lifelong and can be a major impediment to independent living. Therefore, autistic subjects need a long-term educational, psychiatric, and – in selected cases – medical support. It has been previously demonstrated that organized and structured forms of intervention, starting from early childhood and developing during all the different life stages, may improve outcome and quality of life in patients with autism (Howlin et al, 2009). It is therefore conceivable that diverse forms of recovery (e.g. optimal level of motivation, skills, social involvement) may be possible in autism.

There are no fully developed tools with which to evaluate the recovery orientation of a service, but the National Institute for Mental Health in England (NIMHE) has identified the Developing Recovery Enhancing Environments Measure (DREEM) (Ridgway & Press, 2004) as the most promising of an emerging group of recovery sensitive measures. The DREEM permits to collect data on the subjective recovery experience, highlighting the elements that people feel are important to their recovery. Additionally, this questionnaire rates the performance of the mental health service on diverse activities associated with each of these elements. Data from the DREEM may be used not only in evaluating the service but also in educating staff and patients about the recovery, in orienting services towards recovery, in assessing specific recovery oriented programs and supporting on-going quality improvement within the service (Allot et al, 2006).

This study explores the use of DREEM, as a tool to evaluate the effectiveness of recovery-based care in an Italian farm community center specifically designed for adult patients with autism and intellectual disability.

2. Material and methods

2.1. Linguistic validation

The DREEM is a 5-point Likert-type scale that ranges from 1 (Strongly Agree) to 5 (Strongly Disagree). Importantly, lower scores represent higher or more positive ratings and higher scores represent lower or less positive rating. All the questions are stated positively, so no reverse scoring is required. The linguistic validation of the Italian adaptation of the DREEM research tool consisted of three different phases. In the first phase, a board-certified psychiatrist (MB), native speaker of Italian, translated the original instrument into Italian. In the second phase, the Italian version was back translated into English by a professional translator of English background (MT). We compared the original questionnaire and the back translation for coherence and then formulated the initial Italian version of the instrument for patient testing. The third phase, the patient testing panel, was attended by 15 patients recruited in a psychiatric rehabilitation centre in Pavia, Italy. The participants were native speakers of the Italian language. The three steps resulted in the elaboration of the Italian version of the DREEM for the subsequent routine assessment of psychiatric patients or their caregivers.

2.2. Setting

The study was conducted in Cascina Rossago (San Ponzo Semola, Pavia, Italy). Established in 2002, this center is the first Italian farm-community specifically designed for autistic adults (See also Box 1 for more information). The ultimate goal is to improve the growth of each autistic subject in every area of life, using the rural, extended family community as a model. Activities include gardening, animal care, woodworking, carpentry, housekeeping. Additionally, daily schedule presents cognitive activities, stimulating concentration, attention, behavioral control and creative and expressive laboratories, such as music, painting, ceramics. Several sports are performed outside the farm, such as trekking, basket and sailboating. Other form of integration are represented by shopping locally, eating in local restaurants, selling the products of the farm during local festivals. Staff training, supervisions, didactics, meetings are strictly planned within and outside the farm with the special aim to update knowledge and to help care providers in understanding the features of autism and the treatment issues unique to this population. Medical and psychiatric care is assured by the daily presence of a psychiatrist with expertise in the field; this figure supervises activities, programs and staff performance within the farm.

2.3. Participants

The mothers of adult nonverbal subjects with severe autism were invited to take part in the project. The information about the project was given to 24 families at regular parent-caregiver meetings by one of the authors (SU). All patients with autism were recruited from a single farm community center specifically designed for individuals with autism (Cascina Rossago, San Ponzo Semola, Pavia, Italy). The diagnosis in each patient was made on the basis of the Autism Diagnostic Interview-Revised (ADI-R), Italian version (Lord et al, 2003). ADI-R is based on three separate scores. ADI-R domain score A quantifies impairment in social interaction (score range: 0–32), domain score BNV quantifies impairment in nonverbal communication (score range: 0–26), and domain score C quantifies restricted, repetitive, and stereotyped patterns of behavior and interests (score range: 0–16). Higher scores on each indicate worse condition. The cut-off scores of domain score A, domain score BNV, and domain score C are 10, 8, and 3, respectively. A DSM-IV diagnosis of ASD was made for all subjects. All subjects enrolled in the study had a Childhood Autism Rating Scale score > 40. All patients were assessed with Raven's Progressive Matrices, a measure of nonverbal IQ (Raven, 2000). Because of the severity and nonverbality of our patients, the Wechsler Intelligence Scale could not be used. The mothers were required to fill two specific sections (3 and 4) of the DREEM tool in a home-based fashion at two different time points (January 2009 and January 2010). Informed consent was obtained from each mother before examination.

2.4. Statistics

Descriptive statistics, including means and standard deviations were computed for each item of the DREEM. A paired t-test was computed to test the hypothesis that DREEM scores for each item could differ between the two different time points. Statistical significance was set at $p < 0.05$ in all the analyses for which the statistical software SPSS 17.0 version was used.

3. Results

Of the 24 mothers, 17 attended both interviews. Total mean scores in the section Organizational Climate were 1,5 in 2009 and 1,47 in 2010. Total mean scores in the section Recovery Markers were 1,6 both in 2009 and after one year. Statistical analysis indicated no difference for each item between the two different time points.

As depicted in Figure 1 and 2, in both sections and both administrations scores of all items were lower than two.

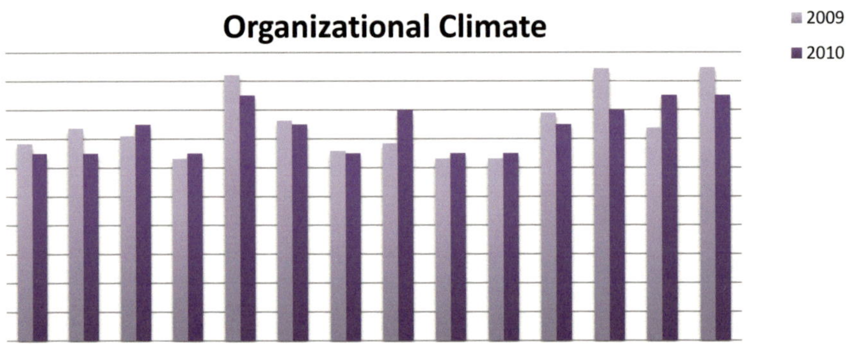

Figure 1. Column graph representing the Organizational Climate in Cascina Rossago

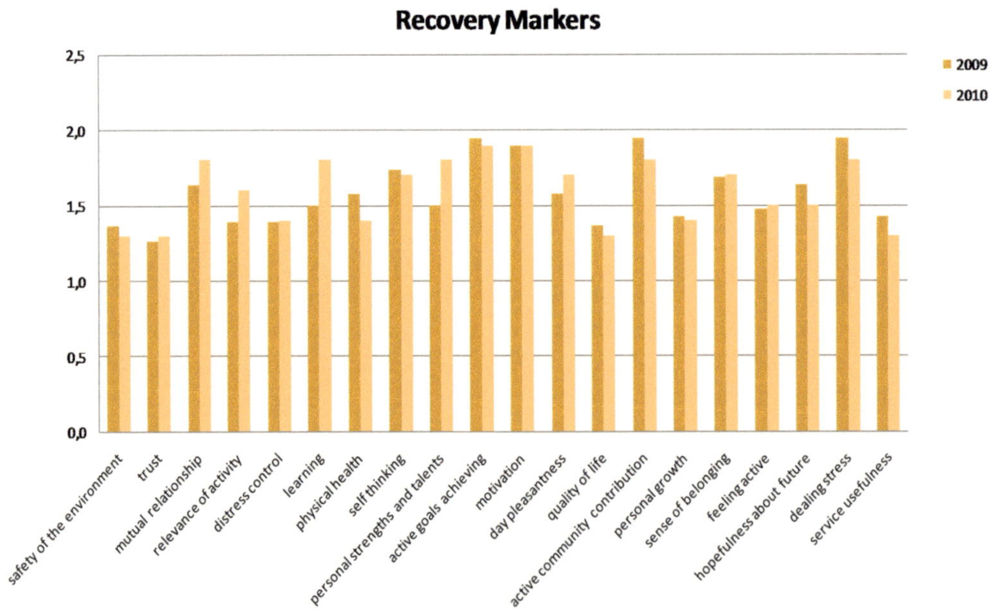

Figure 2. Column graph representing the Recovery Markers in Cascina Rossago

In the section Organizational Climate, the best results, i.e. a score lower or equal to 1.5, are showed in both administrations in the following items: "learning and growth", "hopeful environment", "encouraging service", "friendly service", "respect and esteem", "safety and attractive service", "welcoming staff" (Fig. 1). The item with a score equal or lower than 1.5 in the first administration but not in the second were "relationship" and "feedback relevance", whereas the "creativity of the service" reaches the best score in the second administration (Fig. 1).

In the section Recovery Markers, the best results, i.e. a score lower or equal to 1.5, are showed in both administrations in the following items: "safety of the environment", "trust", "distress control", "quality of life", "personal growth", "feeling active", "service usefulness" (Fig. 2). The item with a score equal or lower than 1.5 in the first administration but not in the second were "mutual relationship", "learning", "personal strengths and talents", whereas "physical health" and "hopefulness about the future" reach the best score in the second administration.

4. Discussion

Our study demonstrates that the section Organizational Climate and Recovery Markers of the DREEM may be useful to evaluate the recovery status in individuals with autism and severe intellectual disability. In fact, the 71% of the mother involved in this study attended both interviews, showing interest in this topic.

The two investigated sections show very good scores at both time points. Particularly, all items had always a score lower than two. Additionally, the absence of difference for each item between the two different time points suggests that a plateau has been reached in our study group.

This results highlight that the farm community is a rehabilitation model promoting the recovery process in autism. The results of both sections reveal that the investigated service is friendly, encouraging, safe and attractive. It favors learning and growth, respect and esteem and is targeted on improving the quality of life of the residents.

This rehabilitation approach, favoring communication, autonomy within a safe and structured framework, seems to be valid in contrasting the core dysfunctions of autism, favoring the growing of each autistic subject in every area of life.

As demonstrated in this work, if the framework is adequate, organized and targeted on patient needs, hopeful and stimulating life is possible also in the severe forms of social and cognitive impairment.

Box 1. General principles of intervention in Cascina Rossago

Constancy and stability are fundamental in planning and doing every activity carried out in Cascina Rossago. Beside these, a constant attention for the core elements of autism is always required, accompanied by a firm organization and permanent education of the staff. Additionally, within the farm it is fundamental to do "real" work, aimed at specific and clear targets.

The rehabilitative approach of Cascina Rossago is based on four essential keywords:

- Ecological approach
- Subjectivity
- Shared problem solving
- Imitation

The ecological approach represents a constant connection between techniques, existential plan, care, organization of the life framework.

A rehabilitative approach targeted on subjectivity implements communication, expression and the ability of making choices, proposing activities fitted on individual motivation and aptitude.

The shared problem solving is probably the pivotal characteristic of Cascina Rossago. It is grounded on Meltzoff's theory "from shared actions to shared minds" (1993), applied to the specific case of subjects with severe social deficit. In an ecological and structured framework, where the relationship is assured by the constant presence of an expert care provider, the autistic subject may be more easily involved in social interaction. Patients and care-providers, together engaged in the activities, share actions, feelings, thoughts and emotions. In this rich context, autistic subjects, overcoming their social difficulties, may more spontaneously detect the "what" and "why" of human intentions.

The last keyword is imitation, necessary to the process of sharing actions and minds, as highlighted by Meltzoff (1993). In the last ten years, diverse theories have postulated an imitation deficit in autism associated with the presence of broken mirror neurons (Williams et al, 2001; Iacoboni & Dapretto, 2006). However, more recently, diverse Authors examined the broken mirror theory of autism concluding that the functioning of the mirror neuron system might be preserved in individuals with ASD to a certain degree (Southgate & Hamilton, 2008; Fan et al, 2010). They highlighted the necessity to study the mirror system within the larger context of the complex circuitries involving imitation, empathy and communication (Arbib, 2007; Southgate and Hamilton, 2009). People with autism show an enhanced automatic imitation effect (Bird et al, 2007). The fact that they can imitate but tend not to do so without instruction suggests that their difficulties arise from problems with knowing when and what to imitate, as a consequence of a reduced sensitivity to social cues. In fact, they can perform a variety of imitation tasks correctly when they are explicitly instructed to imitate (Hamilton et al, 2007).

A rehabilitation approach favoring interaction, communication, autonomy within a safe and structured framework, such as the farm community context, may contrast the core dysfunctions of autism with positive effects the on the whole imitation system. As a consequence, imitation is possible within the farm and may be efficaciously used to favor interaction and communication.

http://dx.doi.org/10.5772/55007

Author details

Marianna Boso[1,2*], Enzo Emanuele[2], Elizabeth Barron[3], Noemi Piaggi[2], Giulia Scanferla[2], Matteo Rocchetti[2], Umberto Provenzani[2], Davide Broglia[2], Paolo Orsi[2], Roberto Colombo[2], Sara Pesenti[2], Marta De Giuli[2], Elena Croci[2], Stefania Ucelli[2], Francesco Barale[2], Jenny Secker[3] and Pierluigi Politi[2]

*Address all correspondence to: marianna_boso@ospedali.pavia.it

1 CPS Pavia, Azienda Ospedaliera Pavia, Pavia, Italy

2 Department of Health Sciences, Section of Psychiatry, University of Pavia, Pavia, Italy

3 Anglia Ruskin University, UK

References

[1] Allot, P, Clark, M, & Slade, M. (2006). Taking DREEM forward: background and summary of experience with REE/DREEM so far and recommendationts. Report prepared for Mental Health Research, Department of Health, available at mentahealtrecovery@blueyonder.co.uk

[2] Anthony, W. A. (1993). Recovery from mental illness: the guiding vision of the mental health system in the 1990s. Psychosocial Reahab. Jour., , 16, 11-23.

[3] Arbib, M. Autism- More than mirror system. Clinical Neuropsychiatry (2007).

[4] Barale, F, & Ucelli, S. (2006). La debolezza piena. Il disturbo autistico dall'infanzia all'età adulta. In Mistura, S. (a cura di) Autismo. L'umanità nascosta, Einaudi 2006.

[5] Emsley, R, Chiliza, B, Asmal, L, & Lehloenya, K. The concepts of remission and recovery in schizophrenia. Curr Opin Psychiatry. (2011). Mar;, 24(2), 114-21.

[6] Fan, Y. T, Decety, J, Yang, C. Y, Liu, J. L, & Cheng, Y. Unbroken mirror neurons in autism spectrum disorders. J Child Psychol Psychiatry. (2010). Sep;, 51(9), 981-8.

[7] Hamilton, A. F, Brindley, R. M, & Frith, U. Imitation and action understanding in autistic spectrum disorders: how valid is the hypothesis of a deficit in the mirror neuron system? Neuropsychologia. (2007). Apr 9;, 45(8), 1859-68.

[8] Howlin, P, Magiati, I, & Charman, T. Systematic review of early intensive behavioral interventions for children with autism. Am J Intellect Dev Disabil. (2009). Jan;, 114(1), 23-41.

[9] Iacoboni, M, & Dapretto, M. The mirror neuron system and the consequences of its dysfunction. Nat Rev Neurosci (2006). , 7, 942-951.

[10] Jacobson, N, & Greenley, D. (2001). What is recovery? A conceptual model and explication. Psychiatr. Serv., , 52, 482-485.

[11] Lord, C, & Rutter, M. LeCouteur, A. ((2003). Autism Diagnostic Interview-Revised (ADI-R). Western Psychological Services.

[12] Meltzoff, A. N. (1993). The centrality of motor coordination and proprioception in social and cognitive development: from shared actions to shared minds. In: The development of Coordination in Infancy, GJP Savelsbergh Ed., 1993 Elsevier Science Publishers, The Netherlands.

[13] Ridgway, P. A, & Press, A. (2004). Assessing the recovery commitment of your mental health service: a users' guide to the Developing Recovery Enhancing Environments Measure (DREEM). (UK ed, Allot P.) UK pilot version (e-mail: mentalhealtrecovery@blueyonder.co.uk).

[14] Ruggeri, M, & Lora, A. Semisa D; SIEP-DIRECT'S Group. The SIEP-DIRECT'S Project on the discrepancy between routine practice and evidence. An outline of main findings and practical implications for the future of community based mental health services. Epidemiol Psichiatr Soc. (2008). Oct-Dec;, 17(4), 358-68.

[15] Southgate, V, & Hamilton, A. F. Unbroken mirrors: challenging a theory of Autism. Trends Cogn Sci. (2008). Jun;, 12(6), 225-9.

[16] Williams, J. H, Whiten, A, Suddendorf, T, & Perrett, D. I. Imitation, mirror neurons and autism. Neurosci Biobehav Rev (2001). , 25, 287-295.

Chapter 7

The Medical Treatment of Autism Disorders

Katarina Dodig-Ćurković, Mario Ćurković and
Josipa Radić

Additional information is available at the end of the chapter

http://dx.doi.org/10.5772/52832

1. Introduction

Autism is one of a group of pervasive developmental disorders and is characterised by qualitative impairments in communication and social interactions and by stereotyped behaviours and interests. Abnormal development is present before the age of 3 years. A quarter of affected children show developmental regression with loss of acquired skills. One third of children with autism have epilepsy and three quarters have mental retardation. Only 15 % of adults with autism lead independent live. Twin and family studies suggest that most cases of autism occur because of combination of genetics factors [1]. Concept of autism has been broadened the last few years from early infantile autism to an autistic spectrum and related communication disorders are grouped together under pervasive developmental disorders or autistic spectrum disorders. People with an autistic disorder have severe difficulties in the integration of perceived in social stimuli into a meaningful entity. More than two people with autistic disorder are also mentally retarded. Autism cannot be cured but adequate intervention can significantly improve the quality of life people with this disorder [2].

The American Psychiatric Association´s last version of the Diagnostic and Statistical manual of Mental Disorders identifies within pervasive developmental disorders five subgroups:

a. autistic disorder,

b. Rett syndrome

c. childhood disintegrative disorder

d. Asperger's disorder and

e. pervasive developmental disorder.

Prenatal exposure to infection and subsequent inflammatory responses have been implicated in the etiology of autism and schizophrenia [3]. Children with autism are vulnerable to anxiety. Also, higher levels of repetative behaviours were associated with more anxiety [4]. Autism is a pervasive developmental disorder characterised by impairment in social interaction and communication, with unusual behaviour [5].

From the previous 4 per 10,000 people, today's prevalence estimates range from 0.6 to around 1% [6].

The onset of autistic disorder is before the age of 3 years and is four to five times more frequent in boys than in girls. Girls with autism are more likely to have more severe mental retardation. The causes of autism spectrum disorders are unknown, although genetic and environmental influence have been implicated. There is increasing evidence that people with autism spectrum disorder have abnormalities in the serotonergic system [7].

There are limited options for pharmacological therapeutic interventions in children with autism disorders and some studies showed that risperidone as an atypical antipsychotic may be in effective for the treatment of people with autism and intellectual disabilities [8].

The primary models of treatment are non pharmacological interventions that include intervention models such as applied behaviour analysis and developmental and structured teaching. The main role of pharmacological interventions is limited to treating symptoms that may be interfering with a child´s ability to learn or function within a particular environment [9].

The prevalence of prescription medications for children with autism is high. Survey indicates that one-half to two-thirds are prescribed at least one medication of any type and about 45% are prescribed at least one psychotropic medication [10-11].

The most commonly prescribed psychotropic medications are antidepressants, stimulants and antipsychotics. The reported prevalence of anticonvulsant medications is approximately 5% among children with autistic disorders and 11-13% among individuals across the life course [12-13].

Atypical neuroleptics have been showed to be useful in the treatment of behavioral symptoms in autism. Attention deficit and hyperactivity disorder medications may be affective for countering the additional features of hyperactivity and short attention span. Antiepileptic drugs for epilepsy and bipolar disorder (Valproat etc.) and selective serotonin reuptake inhibitors have shown promising results for depression [14].

One of the most frequently reported behavioral concerns among children with autism spectrum disorder is high rates of activity and inattention, symptoms that are often associated with attention deficit hyperactivity disorder [15]. The most studied antipshychotic drugs include haloperidol and risperidone [8].

In low dosages, they have been shown to reduce repetitive behaviours (stereotypes) and social withdrawal, as well as a number of related symptoms, such as a hyperactivity, aggression, self-abuse behavior, liability of mood and irritability. All the listed symptoms have appeared in adult patients with schizophrenia, where also with the application of atypical antipsychotics may affect on described clinical picture [16].

70 % of children have mild to moderate learning disability, the remaining 30 % with normal IQ are classified as either high-functioning autism (with language difficulties) or Asperger´s syndrome (with normal language). 1-2 % of those with autism have a normal life; 5 to 20 % of those with autism have a borderline prognosis; but 70 % are totally dependent upon support [17].

The both first and second-generation antipsychotics have shown safety and efficacy in short-term and long-term studies in autism. Safety concerns associated with treatment include the risk of drug-related dyskinesias, which is greater with the first-generation drugs (haloperidol, flufenazin), and the risk of weight gain and associated metabolic problems (increases in glucose and lipids), which is greater with the second generation agents [18-19].

Risperidone has been shown to reduce repetitive behaviours and social withdrawal, hyperactivity, aggression, self-harm behaviour, temper tantrums, lability of mood and irritability [17].

It is also has been proved helpful in treating children and adolescents with autism spectrum disorders behavioural problems, conduct and bipolar disorder, Tourette's syndrome and schizophrenia [20].

Risperidone is a high potency antipsychotic with combined dopamine D2 and serotonin 5-HT2 receptor antagonist properties, has been used to subdue aggressive or self-injurious behaviours. Several reports have suggested that risperidone is effective in diminishing aggressiveness, hyperactivity and self-injurious behaviour in children with autistic disorder. For children with autism, lower dosages ranging from 0.5 to 4 mg per day are generally used [21].

Treatment with atypical antipsychotic olanzapine also can be beneficial in alleviating some behavioral symptoms (irritability, hyperactivity/noncompliance, lethargy), associated with autism [22].

It is very important careful drug titration, usually start with half (½) or one (1) mg in the morning, and then gradually increase the dose in order to prevent adverse effects [23].

It is important to educate parents of children about possible side effects of treatment with antipsychotics. Most often side effects are feeling of stiffness, primarily in the neck and spinal muscles, sedation and weight gain, and in rare cases also galactorrhoea.

This approach can help parents to distinguish between symptoms of disorder and adverse drug effects, reduce their anxiety and intimidation, and simultaneously upgrade compliance and trust between parent and child. These medications are usually studied in adult schizophrenic patients, and there are frequent dilemmas on the use of this group of drugs in children. It is important to always take into account the adverse effects of this group of drugs, time to inform parents about them and have continuous monitoring of children. We have noticed that is often required professional assistance to parents of such children, because of the constant confrontation with difficulties and unpredictable course of illness and fear of the uncertain future of their children.

2. Medical options for children and adolescents

Atypical antypsichotics

Atypical antipsychotics (AAPs) are a group of drugs originally developed to treat psychosis. The group includes compounds brought to the market over the past 10 years as safer and better tolerated alternatives to the existing „typical" antipsychotics. In this group are: *clozapine, risperidone, olanzapine, quetiapine, ziprasidone and aripriprazole.* The target symptoms for pharmacotherapy with AAP typically include aggression, self-injury, property destruction or severe tantrums. Those drugs have lower risk of inducing neurological side effects such as parkinsonism in the short-term and perhaps tardive dyskinesia in the long-term. These newer compounds have been also reported to improve the negative symptoms of schizophrenia (abulia, avolition, flat affect) there is an interest in the notion that this may be relevant to the social withdrawal and lack of spontaneous interaction in autism. The reduced occurence of dyskinesias and the improvement in negative symptoms of schizophrenia may be related to the dual action of five-hydroxytryptamine (5-HT) to dopamine (DA) receptor blockade [24]. The lower *use of clozapine in* autism probably reflects concerns about the risk of blood dysplasia and seizures that are associated with the drug. Additionally, frequent blood tests should be required to monitor for agranulocytosis, which can be challenging in children with autism [25]. *Risperidone* has high affinities for DA D2-D4, 5HT2A, 5-HT2C receptors [26].

It is an atypical antipsychotic (serotonin-dopamine antagonist: second generation antipsychotic) and also mood stabilizers. *Risperidone* blocks dopamine 2 receptors, reducing positive symptoms of psychosis and stabilizing affective symptoms. They also block serotonin 2A receptors causing enhancement of dopamine release in certain brain regions and thus reducing motor side effects and possibly improving cognitive and affective symptoms. Interactions at a myriad of other neurotransmitter receptors may contribute to risperidone's efficacy. Alpha 2 antagonist properties may contribute to antidepressant actions. According to Food and Drug Administration, *risperidone* is commonly prescribed for: schizophrenia ages 13 and older, delaying relapse in schizophrenia, other psychotic disorders, acute mania, mixed mania, autism related irritability in children ages 5 to 16, bipolar depression, behavioral disturbances in children and adolescents and disorders associated with problems with impulse control [27]. Usual dosage range is 2 to 8 mg per day orally for acute psychosis and bipolar disorder, but 0.5 to 2.0 mg per day orally for children and elderly. It is the most frequently used atypical antipsychotic in children and adolescents [28].

Research on *risperidone* shows it to be effective in treating aggressive behaviour in patient population. Also, great improvements have been shown on the verbal communication, apperception and behavioural symptoms [29].

A large number of papers showed that risperidone is a good choice for autistic children with the behavioral problems and irritability. Treatment with all newer antipsychotics was well tolerated with low rates of extrapyramid side effects and serious adverse effects. There is evidence that autistic children and adolescents may be even more prone to weight gain with antipsychotic treatment then patients with general psychiatric conditions. This may occur because these individuals often have less autonomy and ability to take control of appetite, food intake and exercise levels to prevent weight gain [30-31]. Risperidone was found to be more effective than haloperidol in the treatment of behavioral symptoms, impulsivity, language skills, and impaired social relations in children with autistic disorders. Comparing two groups children with autism, there was a greater increase of prolactin in the risperidone group, while alanine aminotransferase had further increased in group that was treated with haloperidol. Also sensory motor behaviours and language at the end of the study showed that sensory motor skills and language subscales scores decreased in the group that was treated with risperidone [32]. With use of risperidone in our Department we achieved reduction of psychomotoric symptoms and reduction of hetero-aggressive and self-destructive behaviours and also improvement in contact with his surroundings. Research on the use of risperidone in the treatment of autistic children in Croatia is rare, given the limited use of risperidone in children younger than 15 years, the question arises about the need to expand the scope of application of risperidone in younger age groups (5).

Many studies show the effectiveness of *haloperidol* in a variety of behavioral symptoms in autistic children. Haloperidol is conventional antipsychotic (neuroleptic, butyrophenone, dopamine 2 antagonist) and according to Food and Drug Administration approved for: manifestation of psychotic disorders, tics and vocal ultrances, in second-line treatment of severe behaviour problems in children of combative, explosive hyperexcitability, second-line short-treatment of hyperactive children, treatment of schizophrenic patients who require prolonged parenteral antipsychotic therapy, bipolar disorder, delirium, behavioral disturbances in dementias. Haloperidol blocks dopamine 2 receptors, reducing positive symptoms of psychosis and possibly combative behaviours. It blocks dopamine 2 receptors in the nigrostriatal pathway, improwing tics and other symptoms in Tourette's syndrome. There is less evidence on the effectiveness of other typical antipsychotics. Haloperidol treatment often causes side effects such as dystonic reactions and dyskinesis. Because of the risk of extrapyramidal symptoms, the use is limited only for refractory cases. The most often seen side effects are: akathisia, neuroleptic-induced syndrome, parkinsonism, tardive dyskinesia, tardive dystonia, galactorrhoea, amenorrhoea, dizziness, sedation, dry mouth, decreased sweating, hypotension, tachycardia, hypertension, weight gain, tardive dystonia. Haloperidol is not intended for use under age 3. Initial oral dose is 0.5 mg/day; target dose 0.05-0.15 mg/kg per day for psychotic disorders and 0.05 to 0.075 mg/kg per day for nonpsychotic disorders [28]. Risperidone is from 2006. year approved by the FDA for the treatment of children older than 5 years. It helps in reducing the symptoms of irritability, aggression and self-harm. The most frequent side effects are weight gain and sedation. Autism is a disorder that lasts a lifetime. Autistic children with IQ higher than 70 and those that develope language communication ability to the fifth or seventh year have the best prognosis. Research on autistic adults have shown that two-thirds of them remain seriously handicapped in complete or partial dependence on caregivers. Only 1-2 percent of persons are capable for independent life, including employment. Five to twenty percent of persons have nearly normal life. All study showed also some adverse events or adverse affects by using risperidone such as: weight gain, somnolence, drowsiness, tremor, dyskinesia, rigidity. Also some authors reported a greater rise in prolactin levels in subjects which were treated by risperidone. These results showed that also it was a greater rise in prolactin levels in subjects in the risperidone compared with those in the placebo althought they did not report clinical events such as gynecomastia or galactorrhoea, that could be elevated prolactin levels [33].

United Kingdom guidelines recommended maximum daily doses of between 2 and 3.5 mg for children weighing under and over 45 kg, respectively. In practice many authorities recommended using very small doses starting at 0.25 mg and increasing very slowly if required.

Aripiprazole is the most recent addition to the list of available AAPs. Studies in adults with schizophrenia have shown it to be an effective antipsychotic with a low risk of side effects and causing reduced levels of weight gain [34]. Aripiprazole is dopamine partial agonist (dopamine stabilizer, atypical antipsychotic, third generation antipsychotic) and commonly prescribed for: schizophrenia ages 13 and older, maintaining stability in schizophrenia, acute mania/mixed mania ages 10 and older, bipolar maintenance, depression, autism-related irritability in children ages 6 to 17, bipolar depresion, other psychotic disorders, disorders associated with problems with impulse control, behavioral disturbances in children and adolescents. Theoretically increases dopamine output when dopamine concentrations are low, thus improving cognitive, negative and mood symptoms. After treatment with aripriprazole, children showed less irritability, hyperactivity, and stereotypes (repetitive, purposeless actions).

Notable side effects must be considered, however, such as weight gain, sedation, drooling, and tremor (35).

By blocking alpha 1 adrenergic receptors it can cause dizziness, sedation and hypotension. Partial agonist actions at dopamine 2 receptors can also cause nausea, occasional vomiting and activating side effects. Mechanism of any possible weight gain is unknown; weight gain is not common with aripriprazole and may have a different mechanism from atypical antipsychotics for which weight gain is common or problematic. Usual dosage range is 15-30 mg/day, but children and patients which not acutely psychotic may need to be dosed lower (2.5 -10 mg /day) in order to avoid akathisia and activation and for maximum tolerability. Aripiprazole was efficacious in children and adolescents with irritability, associated with autistic disorder and was generally safe and well tolerated. Also, it has been shown to be efficacious and generally well tolerated in children and adolescents with schizophrenia and bipolar mania [36-37]. Aripiprazole is approved for use in schizophrenia ages 13 and older, manic/mixed episodes ages 10 and older and irritability associated with autism ages 6 to 17. Clinical experience and early data suggest aripriprazole may be safe and effective for behavioral disturbances in children and adolescents, especially at lower doses. Children and adolescents using aripriprazole may need to be monitored more often than adults and may tolerate lower doses better. Also, may be more risk of weight gain in children than in adults [28]. Data in youth with autism and disruptive behaviour disorders, available only for some antipsychotics, suggest greater weight gain, possibly due to less prior antipsychotic exposure. Metabolic effects differ among second-generation antipsychotics, despite significant weight gain with all studied agents, suggesting additional, weight-independent effects. Pharmacological work indicates that antipsychotics polypharmacy increases the risk for obesity or any other cardiovascular, cerebrovascular or hypertensive adverse event [38].

Serotonin reuptake inhibitors (SRIs)

Repetitive behaviours are a core symptom domain in autism that has been linked to alterations in the serototonin system. While the selective serotonin-receptor inhibitor fluvoxamine has been shown to be effective in adults with autism, as yet no published placebo controlled trials with these agents document safety and efficacy in children with autism. Trials with SRIs suggest benefits in adults with autism spectrum disorders. In the only double-blind study of SRIs in adults with autism to date, McDougle et al. conducted a placebo controled study on fluvoxamine and found that 53% patients were responders compared to placebo group. Significant improvements in repetitive thoughts and behaviour, maldaptive behaviour, aggression, social relatedness and language usage were reported. In contrast to this adult trial of fluvoxamine, a subsequent trial in children and adolescents with autism conducted by the same group found poor response to fluvoxamine, only one of 18 responded, while 14 had adverse effects [39]. To control the depressive symptoms are commonly used fluoxetin, fluvoxamine and sertralin. These act to reduce repetitive, ritualized behaviours and improvement in social skills. SRIs are prescribed for the treatment of co-morbidity associated with autistic spectrum disorders such as depression, anxiety and obsessive-compulsive behaviours [40]. Several studies suggested that children may respond better to low doses of SRIs specifically fluoxetine, than they did to fluvoxamine. Some of the FDA approved drugs used to treat symptoms of autism that can be administrated to children above the age of seven include fluoxetine, fluvoxamine, sertaline and clomipramine. Serotonin reuptake inhibitors (SRIs) such as *clomipramine, fluoxetine, fluvoxamine, sertraline, paroxetine, citalopram and escitalopram that* inhibit the uptake at the presynaptic site.

Fluoxetine is selective serotonin reuptake inhibitor often classified as an antidepressant but it is not only antidepressant. It is commonly prescribed for: major depressive disorder ages 8 and older, obsessive-compulsive disorder ages 7 and older, premenstrual dysphoric disorder, bulimia nervosa, panic disorder, bipolar depression, treatment resistant depression in combination with *olanzapine,* social phobia and posttraumatic stress disorder. Usual dosage range is 20-80 mg for depression and 60 to 80 mg for bulimia. In children it is approved for obsessive-compulsive disorder and it could be helpful in some autistic children with stereotypes reactions. Adolescents often receive adult dose, but doses slightly lower for children. *Fluvoxamine* is also selective serotonin reuptake inhibitor and commonly prescribed for obsessive-compulsive disorder, social anxiety disorder, depression, panic disorder, generalized anxiety disorder, posttraumatic stress disorder and it was helpful for compulsive behaviour and aggression as well as increased prosocial behaviour, in adults with autism, who participated in a double-blind placebo controlled study [41].

In children fluvoxamine is approved for ages 8 to 17 for obsessive-compulsive disorder in inital dose of 25 mg/day at bedtime, increase by 25 mg/day every 4 to 7 days. *Mirtazapine* is an atypical antidepressant in that it posseses both serotoninergic and adrenergic activity.

Study suggested that it can be helpfull for some children with autism for symptoms including aggression, self-injury, irritability, hyperactivity, anxiety, depression and insomia. Adverse effects were minimal and included increased appetite, irritability and transient sedation [42]. Although SRIs may demonstrate therapeutic benefits in autism spectrum disorders, many studies suggest the need for additional randomized controlled trials. Also, given the increased awareness of the dangers associated with SRIs induced activation and agitation, the presence of these side effects in the autistic population warrants closer attention to dosage, titration and subject selection issues [43].

No specific SRIs or dose range has been shown to improve a specific autistic symptom although some patients have demonstrated improvements. Benefits with these drugs in treating functional impairments in autism have been observed. Response to therapy and adverse effects are individualized. Current evidence does not support selection of one SRIs over another for any impairment associated with autism [44].

Tricyclic antidepressants

Autistic spectrum disorders is associated with restricted and/or sterotyped interests or behaviours. Tricyclic antidepressants block noradrenaline and serotonin reuptake. Increasing the availability of these neurotransmitters in the central nervous system. Through their impact on serotonin tricyclic antidepressants have been used in the treatment of autistic symptoms and comorbidities in individuals with autism.

Clomipramine is a tricyclic antidepressant that inhibits the reuptake of both norepinephrine and serotonin. It is commonly prescribed for obsessive-compulsive disorder, depression, severe and treatment-resistant depression, anxiety, insomnia, neuropathic pain/chronic pain. It boosts neurotransmitters serotonin and noradrenaline, blocks serotonin reuptake pump, presumably increasing noradrenergic neurotransmission. Notable side effects are: blurred vision, constipation, urinary retention, increased appetite, dry mouth, nausea, diarrhoea, heartburn, unusual taste in mouth, weight gain, fatigue, weakness, dizziness, sedation, headache, anxiety, nervousness, restlessness, sexual dysfunction, sweating [28]. Clomipramine has been shown to reduce irritability, hyperactivity, inadequate eye contact and inappropriate speech but also further research is required before tricyclic can be recommended for treatment of autistic children [45]. In children and adolescents it is important to monitor patients, particularly during the first several weeks of treatment, also not recommended for use in children age 10.

Anticonvulsants

Anticonvulsants are commonly used in clinical practice in the treatment of autistic children and adults. One in four people with autistic pervasive developmental disorder also have a seizure disorder and usually treated with anticonvulsants such as carbamazepine, lamotrigine, valproic acid. *Lamotrigine* is an anticonvulsant, mood stabilizer, voltage-sensitive sodium channel antagonist and mostly prescribed for: maintenance treatment of bipolar disorder, partial seizures in adults and children age 2 and older, bipolar depression, major depressive disorder, etc. One of the common side effects is rash especially in children ages under 12 and in children taking valproate. Very slow titration may reduce the incidence of skin rush. A study of lamotrigine in 28 children with autism showed no separation between active drug and placebo on measures of stereotypes, lethargy, irritability, hyperactivity, emotional reciprocity, sharing pleasures and in language and communication, socialization and daily living skills noted after 12 weeks [46-47].

The goal of treatment is to alleviate symptoms and improve functioning. If it starts early intensive programs of education and behavioral therapy, can achieve for the child reaches a certain level of independence and gain some social skills. Available pharmacotherapeutic agents are not sufficiently effective for the treatment of the basic symptoms of autistic spectrum. These can help in alleviating comorbid symptoms as a support to the educational, behavioral and cognitive measures. Commonly treated pharmacologically, are the attention deficit, hyperactivity, mood disorders, (anxiety, depression), obsessive-compulsive symptoms, irritability, aggression, propensity for self-harm and sleep disorders. British study from 2004. year by authors Emerson E. and Hatton C, (from Institute for Health Research, Lancaster University, Lancaster UK), conducted on 68 adults whit autism diagnosed before the 1980., with an IQ above 50 is shown following: twelve percent achieved a higher level of independence, ten percent have some friends and are generally employed, but they need support to some extent, nineteen percent have some independence, but generally live at home and they need significant support with the supervision of daily life, forty-six percent require professional, residential care in the specialized institutions for autistic spectrum disorder, with high levels of support and very limited autonomy, while twelve percent have a need for high levels of hospital care.

Electroconvulsive therapy /ECT

ECT is considered as a safe, effective and life-saving treatment in people mainly adults who suffer from affective disorders, acute psychosis and catatonia. There are recent speculations that certain types of autism may be the earliest expression of catatonia and that both disorders have identical risk factors. ECT may improve autism and if started early enough, may prevent further development of autistic symptoms in some children. Researched area that may support the hypothesis that ECT is effective in autism should be pursued [48]. Electroconvulsive therapy should be considered a potentially useful intervention in cases with autistic disorder and a severe comorbid affective disorder.

Acupunture

Acupuncture which involves the use of needles or pressure to specific points on the body, is used widely in Traditional Chinese medicine and increasing within a western medical paradigm. Also, it has been sometimes used as a treatment aimed at improving autism spectrum disorders symptoms and outcomes. Adverse effects noted included bleeding, crying due to fear or pain of life [49].

Neurofeedback

Behaviour therapy improves communications and behavioral functioning in children with autistic symptoms. Neuro-feedback is a noninvasive approach shown to enhance neuroregulation and metabolic function in autistic spectrum disorder [50].

Biological therapies

There are, so called biological therapies ("somewhat controversial") such as secretin, oxytocin, gluten and casein free diet, vitamin B6, magnesium, dimethylglicyn, hyperbaric oxygenation therapy, *melatonin*. Melatonin administration in autistic spectrum disorders is associated with improved sleep parameters, better daytime behaviour and minimal side effects. Commonly used doses are 1-9 mg at night [51]. Vitamin B6 is being tested as a drug to stimulate brain activity. It has been suggested that impairments associated with autism spectrum disorders (ASD) may be partially explained by deficits of omega-3 fatty acids, and that supplementation of these essential fatty acids may lead to improvement of symptoms [52].

Stimulant drugs

Stimulants have been shown to reduce hyperactivity and improve focus, but they may cause behavioral worsening, weight loss and stereotypes de novo. Drugs such as methylphenidate are prescribed for attention deficit hyperactivity syndrome and have proven sufficiently component in treating the similar symptoms of autism. Methylphenidate is stimulant commonly prescribed for narcolpesy and tretament-resistant depression and attention deficit hyperactivity disorder in children ages 6 to 17 and in adults. It increases norepinephrine and dopamine by blocking their reuptake. The main goal of treatment of attention-deficit-hyperactivity is reduction of symptoms of inattentiveness, motor hyperactivity, and or impulsiveness that disrupt social, school and occupational functioning. The common side effects are: insomnia, headache, tics, irritability, anorexia, nausea, abdominal pain, weight loss, blurred vision, and sometimes life-threatening or dangerous side effects as psychotic episodes, seizures, palpitations, tachycardia, hypertension, rare neuroleptic malignant syndrome or cardiovascular abnormalities. Usual dosage range is 2.5 to 10 mg twice per day. Use in young children sholud be reserved for specialist of child and adolescent psychiatry and it is not licensed for children age under 6 years [28].

There is evidence that non-stimulant medication atomoxetine is effective. In both cases adverse events may be increased in this group requiring a slower titration and lower end doses. Atomoxetine is selective norepinephrine reuptake inhibitor commonly prescribed for attention deficit hyperactivity disorder in adults and children over 6 years old. Usual dosage is 0.5 to 1.2 mg/kg per day in children up to 70 kg. Recommended target dose is 1.2 mg/kg per day. It is not licensed for children with structural cardiac abnormalities or other serious cardiac problems [28].

Secretin is a treatment that has received much media attention after reports of efficacy from a small open studies but it controlled studies have failed to show any benefit. In autism also same alternative treatments have been used, but none have shown some benefit [53].

Clonidine is antihypertensive and centrally acting alpha 2 agonist hypotensive agent, nonstimulant for attention deficit - hyperactivity disorder. It is commonly prescribed for attention deficit - hyperactivity disorder, Tourette's syndrome, substance withdrawal, anxiety disorders, menopausal flushing etc. It is not licensed for children and children may be more sensitive to hypertensive effects of withdrawing treatment. Children may be more likely to experience central nervous depression with overdose and might be even exibit signs of toxicity with 0.1 mg of clonidine. Usual dosage is 0.1 to 0.4 mg per day in divided doses. Clonidine is moderately prescribed drug for controlling hypertensive behaviour in autistic children [28]. Other treatment options are: stem cells therapy, sensory-motor therapy, auditory integration therapy, sensory integration therapy and music therapy.

3. Conclusion

The decision to use medications early in the treatment plan for children with autism may be a long-lasting good decision.

Antipsychotics as drugs are intended primarily for the treatment of adult patients with psychotic disorders, but showed the favorable effects on the symptoms in autistic children, especially risperidone. Risperidone has been approved by the US Food and Drug Administration for the symptomatic treatment of irritability (including symptoms of aggression toward others, deliberate self-injuries, temper tantrums, and quickly changing moods) in children and adolescents with autistic disorder.

Some studies suggest that the earlier use of antipsychotics in autistic children may have protective effect on IQ, which further justifies the use of these drugs in early childhood [18]. Research on the use of risperidone in the treatment of autistic disorders among children in Croatia are rare, given the limited use of risperidone in children younger than 15 years, the question arises about the need to expand the scope of application of risperidone in younger age groups.

Also, it may be worth to explore some alternative treatments, such as behavioural interventions, to try to avoid long-term medications. Also it is important to known that although these psychotropic medications have many beneficial effects, they all come with some risk in terms of adverse effects. It is well established that the early detection and treatment of side effects helps reduce their long-term adverse effects on health. Family members are the ones most likely to recognize those reactions such as: weight gain or fatigue. Also it is need for psychoeducational programs for families or individuals with autism that teach parents about medications and the signs and symptoms of potential side effects.

Atypical antipsychotic agents are widely used psyscopharmacological interventions for autism spectrum disorders and among them risperidone has demonstrated considerable benefits in reducing several behavioral symptoms associated with autism spectrum disorders. Althought risperidone has several adverse effects, most are manageable or extremely rare. An exception is rapid weight gain, which is common and can create significant health problems [54]. Aripiprazole, a third generation of atypical antipsychotics, is relatively new drug that has a unique mechanism of action different from other antypshotices. After treatment with aripiprazole children showed less irritability, hyperactivity and stereotypes (repetitive, purposeless actions). Notable side effects must be considered such as weight gain, sedation, drooling and tremor [55].

Autism comprises a clinically heterogeneous group of disorders-named „autism spectrum disorders“ ASD. They share common features or impaired social relationships, impaired language and communications, and repetative behaviours or a narrow range of interests. Management of an autism involves educational, behavioral and medical therapies to promote conversational language and social interactions while mitigating repetitive self-stimulatory behaviours, tantrums, aggression and self-injuries behaviours. Medications, especially atypical antipsychotics can ameliorate specific symptoms such as aggressive or self-injured behaviour. Children treated early can usually be taught, to varying degrees, to communicate, recognize and respond to social interactions, developing imaginative play, and curb all consuming repetative self-stimulatory behaviors [56]. Although most children with autism are healthy, evidence is mounting that medical disorders have a significant effect on behaviours, level of functioning and response to educational therapies. Sensory issues including a blunted pain response, inability to tell others when they are uncomfortable and poor tolerance of medical evaluations can lead to suboptimal medication care. The use of medications has increased as newer medications, especially the atypical antipsychotics, which affect both serotonin and dopamine systems and serotonin reuptake inhibitors (SRIs) which modulate the serotonin system, have been studied in children. In 1997. year the National institute of Mental Health Research Units on Pediatrics Psychopharmacology Autism Network investigated the safety and efficacy of drugs for treating the behaviours associated with autism. Some of conclusions are: no medications are autism specific, marked differences exist in the efficacy of drugs in adults vs. children, individual antipsychotics medications within the same class may differ with respect to their potency and side effect profile, affected individuals may respond differently to the same medication, medication management should be integrated into family centered, multi-modal behavioral and educational program. Typically, first line treatment for children with autism include psychosocial treatments and educational interventions with the goal of maximizing language acquisition, improving social and communications skills and extinguishing of maladaptive behaviours. Currently there are no available standard medication treatments, addressing the core symptoms of autism. There are no pharmacological treatments currently approved by US Food and Drug Administration for autism. When used, pharmacological interventions usually target specific symptoms, accompanying the core symptoms, and severely impairing the individual's functioning, often not allowing for „first line“ educational and behavioral interventions to take place (aggression, self-injurious behavior, compulsive rituals, low frustration tolerance with explosive outbursts, hyperactivity). The newer psychotropics, ecpecially the atypical antipsychotics and the selective serotonin reuptake inhibitors (SSRIs) have more benign side effects profiles than older agents [57]. Despite increased support for pediatric psychotropic use, there is a need for more long-term safety and efficacy studies of existing medications and newer, safer, and more effective agents with fewer side effects for the pharmacological treatment of all childhood disorders in which aggression is prominent. Prescription of antipsychotics drugs requires careful monitoring because of the safety risks and the likelihood of a long-term use.

Drug administration should be initiated at low dosages and subsequent dosage changes sholud be based on tolerability and clinical response. Also children using risperidone may need to be monitored more often than adults.

Author details

Katarina Dodig-Ćurković[1*], Mario Ćurković[2] and Josipa Radić[3]

*Address all correspondence to: katarinadodigcurkovic@gmail.com

1 University Department of Child and Adolescent Psychiatry, University Hospital Center Osijek and Medical faculty in Osijek, Croatia

2 Family medicine Office, Health Center Osijek and Medical faculty in Osijek, Croatia

3 University Depertment of Internal medicine, University Hospital Center, Split and Medical faculty in Split, Croatia

References

[1] Parr, J. (2005). Autism. *Clin Evid*, 14, 275-284.

[2] van Berckelaer-Onnes, IA. (2004). Sixty years of autism. *Ned Tijdschr Geneeskd*, 148(21), 1024-1030.

[3] Mayer, U., Feldon, J., & Dammann, O. (2011). Schizophrenia and autism: both shared and disorder-special pathogenesis via perinatal inflammation. *Pediatr Res*.

[4] Rodgers, J., Riby, D. M., Janes, E., Connolly, B., & Mc Conachie, H. (2011). Anxiety and repetative behaviours in autism spectrum disorders and wiliams sy. A Cross-syndrome comparasion. *J Autism Dev Disord*.

[5] Dodig-Curkovic, Curkovic. M., Radic, J., & Radić, M. (2011). The treatment of autistic children with risperidone. *Coll Antropol*, 35(1), 298-301.

[6] Stanković, M., Lakić, A., & Ilić, N. (2012). Autism and autistic spectrum disorders in the context of new DSM-V classification and clinical and epidemiological data. *Srp Arh Celok Lek*, 140(3-4), 236-43.

[7] Muhler, R., Trentacoste, S. V., & Rapin, I. (2004). The genetics of autism. *Pediatrics*, 113, 472.

[8] Malone, R. P., & Waheed, A. (2009). The role of antipsychotics in the menangment of behavioral symptoms in children and adolescents with autism. *Drugs*, 69(5), 535-48.

[9] Allesandri, M., Thorp, D., Mundy, P., & Tuchman, R. F. (2005). Can we cure autism? From outcome to intervention. *Rev Neurol*, 15(40), S 131-6.

[10] Aman, M. G., Lam, K. S. L., & Van Bourgondien, M. E. (2005). Medication patterns in patients with autism:Temporal, regional and demographic influences. *Journal of child and adolescent Psychopharmacology*, 15, 116-126.

[11] Martin, A., Scahill, L., Klin, A., & Volkmar, F. R. (1999). Higher-functioning pervasive developmental disorders: Rates and patterns of psychotropic drugs use. *Journal of the American Academy of Child and Adolescent Psychiatry*, 38, 923-931.

[12] Langworthy-lam, K. S., Aman, M. G., & Van Bourgondinen, M. E. (2002). Prevalence and patterns of used psychoactive medicines in individuals with autism in the Autism Society of North Caroline. *Journal of Child and Adolescent Psychopharmacology*, 12, 311-321.

[13] Witwer, A., & Lecavalier, L. (2005). Treatment incidence and patterns in children and adolescents with autism spectrum disorders. *Journal of Child and Adolescent Psychopharmacology*, 15, 671-681.

[14] Benvenuto, A., Battan, B., porfirio, M. C., & Curatolo, P. (2012). Pharmacotherapy of autism spectrum disorders. *Brain Dev.*

[15] Handen, B. L., Taylor, J., & Tumuluru, R. (2011). Psychopharmacological treatment of ADHD symptoms in children with autism spectrum disorder. *Int J Adolesc Med health*, 23(3), 167-73.

[16] Malone, R. P., Gratz, S. S., Delaney, M. A., & Hyman, S. B. (2005). Advances in drug treatments for children and adolescents with autism and other pervasive developmental disorders. *CNS Drugs*, 19(11), 923-34.

[17] Sample, D., Smyth, R., Burnd, J., Darjee, R., & Mc Intosh, A. (2005). Oxford Handbook of psychiatry. *New York:Oxford University Press.*

[18] Begovac, I., Begovac, B., Majić, G., & Vidović, V. (2009). longitudinal studies of IQ stability in children with childhood autism-literature survey. *Psychiatr Danub*, 21(3), 310-9.

[19] Malone, R. P., Delaney, M. A., Hyman, S. B., & Cater, J. R. (2007). *J Child Adolesc Psychopharmacol*, 17(6), 779-90.

[20] Chevreuil, C., Reymann, J. M., Freumax, T., Polard, E., Seveno, T., & Bentue-Ferrer, D. (2009). Risperidone use in child and adolescent psychiatric patients. *Therapie*, 63, 359.

[21] Sadock, J. K., & Sadock, A. V. (2008). Pervasive developmental disorders. Autistic disorder. *In: SADOCK JK, SADOCK AV Synopsis of psychiatry (Lippincott Williams and Wilkins, Philadelphia.*

[22] Fido-Saad, A. A. (2008). Olanzapine in the treatment of behavioral problems associated with autism: an open-lebel trial in Kuwait. *Med Princ Prac*, 17, 415.

[23] Resenbaum, J. F., Arana, G. W., & Hyman, E. S. (2005). Antipyschotics. *In: Rosenbaum JF, Arana GW, Hyman ES, Handbook of Psychiatric Drug Therapy (Lippincott Williams & Wilkins, Philadelphia.*

[24] Anderson, L. T., Campbell, M., Adams, P., Small, A. M., Perry, R., & Shell, J. (1998). The effects of haloperidol on discrimination learning and behavioral ymptoms in autistic children. *J Autism disord*, 227-39.

[25] Chen, N. C., Bedair, H. S., Mc Kay, B., Bowers, M. B., & Mazure, C. (2001). Clozapine in the treatment of agression in a adolescent with autistic disorder. *J Clin psychiatry*, 62(6), 479-80.

[26] Leysen, J. E., Gommeron, W., Eens, A., de Chaffoy de, Courcelles. D., Stoof, J. C., & Janssen, P. A. (1988). Biochemical profile of risperidone, a new antypsihotics. *J Pharmacol Exp Ther*, 247(2), 661-70.

[27] Lieberman, J. A., Stroup, T. S., & Mc Evoy, J. P. (2005). Effectiveness of antipsychotic drugs in patients with chronic schizophrenia. *N. Eng J Med*, 353(12), 1209-23.

[28] Stahl, M. Stephen. (2011). The Prescriber Guide, Stahls Essential Psychopharmacology. Fourth Edition. *Cambridge University Press. Cambridge.*

[29] Findling, R. L. (2008). Atypical antipschotic treatment of disruptive behavior disorders in children and adolecents. *J Clin Psychiatry* [69, 4].

[30] De Hart, M., Dobbelaere, M., Sheridan, E. M., Cohen, D., & Correll, C. U. (2011). Metabolic and endocrine adverse effects of second-generation antipsychotics in children and adolescent: A systematic review of randomized, placebo-controlled trials and guidelines for clinical practice. *European Psychiatry*, 26(3), 144-148.

[31] Hellings, J. A., Zarcone, J. R., Reese, R. M., Valdovinos, M. G., Marquis, J. G., Fleming, K. K., & Schroeder, S. R. (2006). A Crossover Study of risperidone in children,. adolescents and adults in mental retardation. *Journal of Autism and developmental disorders*, 36(3), 401-411.

[32] Miral, S., Gencer, O., Inal-Emiroqlu, F. N., Baykara, B., Baykara, A., & Dirik, E. (2008). Risperidone versus haloperidol in children and adolescents with AD: a randomized, controlled, double-blind trial. *Eur Child Adolesc Psychiatry*, 17(1).

[33] Mc Pheeters, M., Warren, Z., Sathe, N., & Bruzek, J. (2011). A systematic review of medical treatments for children with autism spectrum disorders. *Pediatrics*, 127.

[34] Kane, J. M., Carson, W. H., Saha, A. R., Mcquade, R. D., Ingenito, G. G., Zimbroff, D. L., & Ali, M. W. (2002). Efficiacy and safety of aripriprazol and haloperidol versus placebo in patiens with schizophrenia and schizoaffective disorder. *J clin Psychiatry*, 63(9), 763-71.

[35] Ching, H., & Pringsheim, T. (2012). Aripiprazole for autism spectrum disorders (ASD). *Cochrane Database Syst Rev*, 16, 5.

[36] Owen, R., Sikich, L., Marcus, N. R., Lislie-Corey, P., manos, G., Mc Quade, D. R., Carson, W. H., & Findling, L. R. (2009). Aripriprazole in the treatment of irritability in children and adolescents with autistic disorder. *Pediatrics*, 124(6), 1533-40.

[37] Marcus, R. N., Mc Quade, R. D., Carson, W. H., et al. (2008). The efficacy and safety of aripriprazole as adjunctive therapy in major depressive disorder: a second multicenter, randomized, double-nlind, placebo-controlled study. *J Clin Psychopharmacol*, 28(2), 156-65.

[38] Mayyan, L., & Corell, C. U. (2011). Weight gain and metabolic risk associated with antipsychotic medications in children and adolescents. *J Child adolesc Psychopharmacol*, 21(6), 517-35.

[39] Mc Dougle, C. J., Kresch, L. E., & Posey, D. J. (2000). Repetitive thoughts and behaviour in pervasiveive developmental disorders:treatment with serotonin reuptake inhibitors. *J Autism Dev Disord*, 30, 427-435.

[40] Williams, K., Wheeler, D. M., Silove, N., & Hazell, P. (2010). Selective serotonin reuptake inhibitors for autism spectrum disorders. *Cohrane Database Syst Rev* [8].

[41] Mc Doughle, C. J., Naylor, S. T., Cohen, D. J., Volkmar, F. R., Heninger, G. R., & Price, H. L. (1996). A double-blind, placebo controlled study of fluvoxamine in adults with autistic disorder. *Arch Gen Psychiatry*, 53(11), 1001-8.

[42] Posey, D. J., Genin, K. D. K. O., Hn, A. E., Swiezy, N. B., & Mc Dougle, C. J. (2001). A naturalistic open-label study of mirtazapine in autistic disorder and other pervasive developmental disorders. *J Child Adolesc Psychopharmacolol*, 11(3), 267-77.

[43] Kolevzon, A., Mathewson, K. A., & Hollander, E. (2006). Selective serotonin reuptake inhibitors in autism: a review of efficacy and tolerability. *J Clin Psychiatry*, 67(3), 407-14.

[44] Moore, M. L., Eichner, S. F., & Jones, J. R. (2004). Treating functional impairment of autrism with selective serotonin-reuptake inhibitors. *Ann Pharmacother*, 38(9), 1515-9.

[45] Hurwitz, R., Blackmore, R., Hazell, P., Williams, K., & Woolfenden, S. (2012). Tricyclic antidepressants for autism spectrum disorders in children and adolescents. *Cohrane database syst Rev*, 3.

[46] Hollander, E., Dolgoff-kaspar, R., Cartwright, C., Rawitt, R., & Novotny, S. (2001). An open trial of divalproex sodium in autiosm spectrum disorders. *J Clin Psychiatry*, 62(7), 530-4.

[47] Belsito, K. M., Law, P. A., Kirk, K. S. A., Landa, R. J., & Zimmerman, A. W. (2001). Lamotrigine therapy for autistic disorder: a randomized, double-blind, placebo-controlled trial. *J Autism Dev Disord*, 318(2), 175-81.

[48] Dhossche, D. M., & Stanfill, S. (2004). Colud ECT be effective in autism? *Medical Hypothesis*, 63(3), 371-6.

[49] Cheuk, D. K., Wong, V., & Chen, W. X. (2011). Acupuncture for autism spectrum disorders (ASD). *Cohrane Database Syst Rev*, 9.

[50] Coben, R., Linden, M., & Myers, T. E. (2010). Neurofeedback for autistic spectrum sidorder: a review of the literature. *Appl Psychophysiol Biofeedback*, 35(1), 83-105.

[51] Rossignol, D. A., & Frye, R. E. (2011). Melatonin in autism spectrum disorders: a systematic review and meta-analysis. *Dev Med Child Neurol*, 5389, 783-92.

[52] James, S., Montgomery, P., & Williams, K. (2011). Omega-3 fatty acids supplementation for autism spectrum disorders (ASD). *Cohrane Datebase syst Review* [9].

[53] Malone, R. P., Gratz, S. S., Delaney, M. A., & Hyman, S. B. (2005). Advances in drug treatments for children and adolescents with autism and other pervasive developmental disorders. *CNS Drugs*, 198(11), 923-34.

[54] Sharma, A., & Shaw, S. R. (2012). *J Pediatr Health Care*, 26(4), 291-9.

[55] Ching, H., & Pringsheim, T. (2012). Aripriprazole for autism spectrum disorders (ASD). *Cohrane Database Syst Rev*, 16, 5.

[56] Miles, J. H., Mc Cathren, R., Sticher, J., & Shinawi, MD. (1993). Autism spectrum disorders. *GeneReviews, NCBI Bookshelf. Pagon RA, Bird TD Dolan CR et al editors. GeneRewievs. Seattle, University of Washington; Seattle.*

[57] Nikolov, R., Jonker, J., & Scahill, L. (2006). Autistic disorder: current psychopharmacological treatments and areas of interes for future developments. *Revista Bras Psychiatr*, 28(1).

Section 5

The Law and Autism

Chapter 8

Forensic Issues in Autism Spectrum Disorder: Learning from Court Decisions

Ian Freckelton

Additional information is available at the end of the chapter

http://dx.doi.org/10.5772/55400

1. Introduction

Utilising a series of recent Australian decisions, this paper reviews contemporary approaches of the courts in relation to the relevance of Autism Spectrum Disorder (ASD) to criminal responsibility and culpability. It employs the definition of the disorder expected to be employed in DSM-V (American Psychiatric Association, 26 January 2011), recognising that the new formulation is broad-brush and does not avail itself of the nomenclature or separate diagnostic criteria for Asperger's Disorder.

It identifies a significant level of unawareness in Australian courts as to the distinctiveness, and more importantly the potential significance of the distinctiveness, of the world experience of those with symptomatology of the disorder and explores the challenge for mental health experts in effectively assisting judicial officers to understand the internal experience of those with ASD and differentiating it from personality disorders and psychopathy (see though Fitzgerald, 2010).

Drawing on the reasoning in recent decisions, it argues that, while Asperger's Disorder, and to a lesser degree ASD, have now penetrated the public and judicial consciousness to some degree, there is a significant forensic distance to travel and many challenges to be overcome before courts are enabled meaningfully to appreciate for any given defendant the impact likely to have been exercised on offending behaviour by an ASD. Identifying something of a backlash against such a disorder being viewed as significantly mitigating in Australia, it advances proposals for how mental health professionals can sensitise the courts more informedly to the potential forensic significance of ASD.

2. ASD and the law

The preponderance of scholarship in relation to ASD and the criminal law has related to persons with Asperger's Disorder (see generally Murrie et al, 2002; Silva, Ferrari and Leong, 2003; Barry-Walsh and Mullen, 2004; Haskins and Silva, 2006; Warren, 2006; Langstrom et al, 2009; Freckelton and List, 2009; Browning and Caulfield, 2011; Freckelton, 2011; Freckelton, 2013). It has identified issues that have arisen in relation to fitness to be interviewed, fitness to stand trial, capacity to form intent, fitness for extradition, defences, such as self-defence, insanity/mental impairment, diminished responsibility, and, in particular, sentencing (see Warren, 2006; Freckelton and List, 2009; Freckelton, 2013). It has also analysed particular categories of offences that have been committed by persons with ASD: offences of violence, offences against public order, sexual offences, arson offences and computer offences (see Realmuto and Ruble, 1998; Milton et al, 2002; Mouridsen et al, 2008; Freckelton, 2011; 2013;). Awareness amongst judicial officers' decisions of the potential relevance of the disorder arises principally from expert insights into the fact that those with ASD experience the world in a way that is at major variance from those without the disorder. The disorder can be marked by obsessionality, inability to apprehend verbal and non-verbal cues, lack of empathy, rigidity, literalism in response, naivete and a propensity to panic and behave impulsively and unpredictably in unfamiliar environments. (see eg *R v Mueller*, 2005 at [92]; *McC v The Queen*, 2007). Such persons may also be very suggestible (see eg *IA v The Queen*, 2005 at [8]), have an aversion to being touched by others (see eg *ZH v The Commissioner of Police for the Metropolis*, 2012) or be distressed by sensory perceptions such as noise or intense light (see Cascio et al, 2012), possibly because of sensory processing deficits (see Lane, Young, Baker and Angley, 2010). There are important indications that persons with ASD are significantly over-represented as persons in custody (Cashin and Newman, 2009; Mayes, 2003; Scragg and Shah, 1994).

There are indications arising from court decisions that mental health experts in the forensic area and judicial officers alike have not always been as aware of the counter-intuitive characteristics of ASD as they need to be so that fair decisions are made by courts about procedural issues arising in criminal trials, as well as about both the criminal responsibility and culpability of those with ASD.

As of 2013, it can be said that the incidence of cases in which ASD is invoked as relevant to criminal trials or sentencing appears to be growing. Part of this is attributable to increasing awareness of ASD within the general population and by extension within the legal community. This highlights the need for more focused and expert specialist assessment of defendants for whom their ASD may constitute an important aspect of their defence (Freckelton and Selby, 2013; Freckelton, 2013). In this paper recent Australian decisions have been selected for evaluation as illustrative of the complex forensic and mental health issues which continue to be confronted by the courts, even in a jurisdiction which has had a disproportionate number of appellate decisions which have wrestled with the forensic significance of ASD.

3. R v George (2004)

One of Australia's earliest and most significant decisions in relation to ASD in the criminal law, but which has not previously been scrutinised, save in relation to issues relating to hoarding and the law (Freckelton, 2012), was that of *R v George* (2004).

George was found guilty at first instance in the New South Wales Supreme Court of manslaughter of his 86 year old mother (for whom he was the primary carer) by criminal negligence arising from his failure to provide her with proper nutrition, hydration, medication and medical attention. He was sentenced after trial to imprisonment for seven years with a non-parole period of four years.

George appealed to the New South Wales Court of Appeal (*R v George*, 2004) on the ground that insufficient weight was given by the sentencing judge to the fact that he suffered from Asperger's Disorder and other psychological issues that resulted from a solitary life and social dysfunction. The Court of Appeal allowed his appeal and reduced his sentence to three years and six months' imprisonment with a non-parole period of two years.

The evidence before the Court was that at the time of his mother's death George was 58 years of age and had never married. He lived until the time of her death with his mother and a developmentally delayed sister. He had generally been unemployed although latterly he had undertaken some minor administrative functions at a chiropractic clinic.

About two years before her death, George's mother had instructed her children not to arrange home care for her as she was embarrassed by the state of her house and its surrounds. The garden was seriously overgrown and unkempt and the interior of the house was overrun by papers, some of which were kept in bags and piled in various rooms. The shower and the bath had not worked in some time. Newspapers were stacked in the shower recess and the toilet leaked. Thick dust and cobwebs were to be found throughout the house.

Evidence before the Supreme Court at the time of the manslaughter trial established that George's mother had been a domineering person and had vigorously resisted all efforts to take her to hospital and provide her with home care. She had an aversion to being showered. George's culpability depended on the sufficiency or otherwise of efforts he had made to provide her with care as her condition deteriorated. When ambulance officers discovered her, she was bed ridden and covered in sores. She was wearing soiled clothing and her bed and person were covered in human vomitus, faeces, urine and body fluids. She was severely malnourished and shortly afterwards died of bronchial pneumonia. Evidence established that she had not been provided with prescribed medicine for some years and that she had suffered significantly, especially in the latter stages of her life.

Psychiatric evidence shed some light on the reasons for the insufficiency of George's responses to his mother's worsening health condition. It suggested that George had a "mild variant of an autistic disorder", a diagnostic criterion for which was "a lack of social or emotional reciprocity, which could partly explain his apparent lack of concern for his mother's condition." (*R v George*, 2004: [23]) The psychiatrist noted that George had an apparently consuming interest (typical of persons with Asperger's Disorder) in all things related to railways. He noted that: "The idiosyncratic thinking that is usually observed in the presence of Mr George's disorder could also explain his rather literal interpretation of his mother's instructions and his apparent lack of concern when interviewed about the events." (*R v George*, 2004: [23]) He explained this further as incorporating a difficulty in George having a normal level of empathy and an capacity to recognize and respond to the reactions of others by reason of his lack of empathy.

On appeal the Court found that the sentencing judge's failure to deal explicitly with these considerations "was a deficiency of some importance" in that it significantly reduced George's level of culpability:

Upon the evidence his capacity to respond to his responsibilities was clearly impaired by an unusual personality disorder arising from his history of social dysfunction, as evidenced by the utterly bizarre circumstances in which he, and the immediate family, lived.

The case is a tragic and wholly exceptional one, and we are driven to the conclusion that the Applicant's objective criminality was overstated by his Honour. As Dr Nielssen explained, Asperger's Syndrome is not normally associated with criminal offending, and the risk of the Applicant reoffending, or of being placed in a similar situation, is minimal. Personal deterrence is, accordingly, of little relevance. (*R v George*, 2004: [42]-[43])

R v George (2004) is significant as it is Australia's first appellate authority on the potential relevance of ASD symptomatology to the evaluation of criminal culpability. Its recognition of the distinctive propensity of persons with an ASD to comply with the plain words of instructions from a trusted person and not to respond emotionally even to manifest suffering, if such a response is interpreted by them as inconsistent with the instructions given to them, is consistent with clinical insights into the characteristics of persons with ASD. It highlights that the lack of empathy of a person with ASD is prone to be interpreted as malign intent (eg criminal mens rea) unless counter-intuitive expert evidence is provided to the contrary.

4. R v Hampson (2011)

Bradley Hampson, a man of 29, pleaded guilty to a range of offences relating to possession and distribution of child pornography. He was sentenced to three years' imprisonment with release after 12 months and then two years' probation. He appealed to the Queensland Court of Appeal (*R v Hampson*, 2011) on the basis that his sentence was manifestly excessive. He had a prior conviction for using a carriage service to menace, harass or cause offence – when he had telephoned persons and made lewd inquiries of them.

Evidence before the Court of Appeal established that Hampson had been diagnosed with autism by a psychologist to whom he had been referred when he applied for a disability support pension. The criminal conduct the subject of his sentence appeal included posting sexually offensive observations on the tribute page for a murder victim and the distribution of obscene comments and sexualized depictions of children. The sentencing judge described Hampson's conduct as "depraved".

The sentencing judge acknowledged that: "It seems that the origin of your offending may lie somewhere in your history of autism and in your own social ineptitude which led you to misusing the internet in the way you did." (*R v Hampson*, 2011: [57]) The Court of Appeal, however, concluded that little weight should be given to Hampson's autism for the purposes of assessing his criminal culpability and that the trial judge's incorporation of the condition into his sentencing analysis was proper but sufficient.

The *Hampson* decision raises the difficult issue of the blameworthiness of persons with ASD for conduct that would be considered difficult to understand and repugnant in ordinary members of the community. Persons with ASD are frequently absorbed by persons, objects and details. They can be very concrete in their reactions (see eg *R v George*, 2004). They can have a propensity to engage in repetitive and obsessive behaviours, especially within the unthreatening environment of the Internet.

In the Ontario case of *R v Somogyi* (2011) a man found guilty of luring two girls under the age of 14 and inviting them to engage in sexual touching was sentenced "only" to a conditional sentence with house arrest in part because of his having ASD. In taking what he acknowledged was an unusual step Anderson J observed that Mr Somogyi had the social age of a 12 year old: "This was a fantasy world for Mr Somogyi, where he could communicate with children that were perhaps closer to his own emotional age. It appeared clear to me that part of the communication with the two undercover officers was sexual in nature, but that part of the communication was as friends, consisting of sharing music, pictures and conversation. In this communication, Mr Somogyi could be the knowledgable outgoing leader, not the shy awkward adult." (at [36])

The environment of the Internet can enable persons with ASD to act out their sexual attractions and impulses, as well as feelings of distress and anger, without the confronting exigencies of direct person-to-person interaction. When such a pattern of fascination is coupled with a reduced level of socialization and capacity for understanding of and empathy with others' sensibilities, there is the potential for them to engage in conduct that is alienating and frightening but whose resonances and consequences are not (well) understood by them. Expert mental health evidence to explicate these limitations on the part of persons with ASD is extremely important if unfair harshness in sentencing is to be avoided.

5. DPP v HPW (2011)

In *Director of Public Prosecutions v HPW* (2011) the Victorian Court of Appeal heard an appeal brought by the prosecution contending that the sentencing judge at first instance had wrongly found a causal connection between HPW's Asperger's Disorder and his sexual offending, had erred in imposing a manifestly inadequate sentence and had inadequately cumulated the penalties he imposed for a significant number of sex offences.

HPW was found guilty at first instance of eight charges, three of which were representative of many instances of sex offending, committed against his biological daughter during a time when she was aged 11 and 12. They involved multiple instances of oral, digital and anal penetration as well as instances of masturbation and of encouraging the family dog to lick his daughter's vagina.

When interviewed by the police, HPW admitted sodomising his daughter and explained that it was "just as an experiment". He said by way of explanation that "it was just sexual gratification for myself" and commented that he was "probably a psycho".

HPW was aged 47 at the time of sentencing and without prior convictions. He had served a lengthy period of time in the army until he was discharged in 2007 for not handing back some hand grenades. He had two children from a marriage that lasted over a decade, after which he formed a relationship that involved bestiality and anal sex with another woman.

A psychologist who examined him formed the view that the offending with his daughter stopped when "he realised what he was doing". HPW's Asperger's Disorder went undiagnosed until after the criminal charges were laid. A psychologist who assessed him expressed the view (at [37]) that he had

---

significant deficits in social interaction; restricted behaviour, interests and activities; clinically significant impairment in social or other important areas of functioning; no apparent language impairment; and no apparent cognitive impairment. He is somewhat atypical in his awareness of his deficiencies in empathy and friendship skills.

Another psychologist, Dr Kennedy, whose report was tendered at the plea hearing, stated (at [47]):

In this case, victim empathy should be commented on for specific reasons, particularly in relation to [HPW]'s cognitive distortion associated with the offences. In this matter, he has reported that while carrying out the sexual offences he considered that [his daughter] *was experiencing the sexual abuse in a matter-of-fact way* as if the activities were normal, and nothing more than her daily activities.

Discussion of this issue occurred at some length. I should note that [HPW] did not appear to be attempting to minimise this behaviour in this [sic], but *was attempting to explain how he saw [his daughter's] response to the sexual abuse*. He thought at the time for her, it was "something to do... as if it was an activity such as playing cards or watching TV" that *had no impact on her at an emotional level*. When asked about his understanding of the effects of the sexual abuse on [his daughter], he reported in a very distinct way that the impact has been "huge... I think I've ruined her... she'll never be able to see me in the same light... it will be very difficult for her with partners in the future".

He added (at [50]):

[There is a] focus on deficient empathy, which is clearly relevant in this case, interpersonal naivety which appears to be the case in this matter, sexual frustration which is clearly relevant in this case, and immediate confession, which from my understanding, is also present. Additionally, there are sexual preoccupations, which do appear relevant in this case.

Dr Kennedy expressed the view that at the time of his offending HPW was unaware of the distress he was causing to his daughter but contended that since that time, with professional assistance, he had acquired genuine empathy and remorse. He observed that there had been no grooming process, as is often seen in sex offending cases.

The Court of Appeal found that the evidence of the expert gave no support for the foundation of the plea made on HPW's behalf, and which was (wrongly) accepted by the sentencing judge, that HPW misread his daughter's behaviour as providing encouragement to him by hints or signals, to engage in the sexual offending. Justice Tate (at [53]), writing the leading judgment, found that Dr Kennedy's opinion:

---

suggested that the sexual offending occurred in a context in which (1) the respondent had sexual preoccupations with his daughter, fantasising about her in a manner reflective of his previous unusual sexual relationship with an earlier partner of whom his daughter reminded him; (2) he was sexually frustrated with his current partner; (3) his level of alcohol abuse led to disinhibition; and (4) his deficient empathy meant that he believed that his sexual offending was having no emotional impact on his daughter. Dr Kennedy's opinion did not provide a proper evidentiary base supporting the finding of the sentencing judge that the respondent 'may have misinterpreted [his] daughter's cues'.

She found that the plea by counsel misrepresented the expert report. To the extent that Dr Kennedy had commented "it is highly likely that [HPW's] behaviour is best explained by the presence of an Autism Spectrum Disorder", Tate JA found that this could not support the proposition that there was a causal connection between his conduct and his misreading of his daughter's behavioural cues. This led Tate JA (with Neave and Mandie JJA agreeing) to find a sentencing error to have been committed at trial. They also found that HPW's Asperger's Disorder should not have led to a significant moderation in the sentence imposed upon him, and that his sentence was not sufficiently cumulated to reflect the "debased and humiliating nature of the offending, the core breach of trust, or the effect of the offending upon [HPW's] daughter" (at [82]).

While the court did not find that HPW's Asperger's Disorder reduced his moral blameworthiness for the purposes of sentencing, it did accept that it was appropriate to view his disorder as a mitigating factor to the extent that it was likely to make his service of a custodial sentence more burdensome for him. It ordered his sentence to be increased from seven and a half years' imprisonment with a non-parole period of five years and six months to nine years and six months' imprisonment with a non-parole period of six years and six months.

The *HPW* decision is a salutary reminder that judges' evaluation of the relevance of Asperger's Disorder will vary from case to case, depending upon factors such as the severity of the disorder, the nature of the offending and the proven relationship between the disorder and the particular offending. On some occasions it will be regarded by sentencers as powerfully relevant, while on others it may be found nearly irrelevant. This is consistent with the position with psychiatric disorders that more commonly intersect with the cases determined by the criminal courts – for instance, the fact alone that a person satisfies the DSM criteria for schizophrenia does not of itself relieve the person of criminal responsibility or culpability. However, more can be observed. A real question will arise on occasions about the extent in a meaningful sense that a person with ASD will be aware, other than at a superficial or, to them, a theoretical level, of the wrongfulness of their behaviour and of the consequences that it is likely to bring for their victim. In such a situation, real questions arise in relation to their criminal culpability and therefore the basis upon which they should be sentenced. This issue is at its most confronting when, as in the *HPW case,* the conduct is particularly unpleasant.

6. R v Sokaluk (2012)

Brendan Sokaluk was convicted by a jury in Melbourne, Australia, of ten counts of arson causing death, an offence carrying a maximum sentence of 25 years of imprisonment for each charge. The sentencing judge, Justice Coghlan of the Supreme Court of Victoria, found that Sokaluk had intentionally lit a fire in eucalypt plantations and in two other places, knowing that his actions would cause damage, and in fact causing the deaths of ten people. He accepted that Sokaluk did not intend to cause loss of life but found that, nonetheless, the fact was that he did cause multiple deaths on a day that became known in Australia as "Black Saturday", when strong winds built up and temperatures exceeded 46 degrees Celsius.

Evidence before the jury showed that when he was being evacuated from the fire zone Sokaluk told a lie because his father earlier in the day had advised him not to go to the hills. This was but one of a number of untruths told by Sokaluk and which raised the issue of whether he understood full well that his actions were wrongful and could lead to adverse consequences. Another occurred on the next day when he returned to the scene of one of the fires and saw that his car had been destroyed by the fire. Within an hour and a half he made a claim on his insurance policy. Justice Coghlan found that his level of functioning during the call "was at the very least reasonable" (at [23]). He drew adverse inferences for Sokaluk's level of understanding and functioning from this conduct.

In the next days Sokaluk made various comments about who had been responsible for lighting the fire and then on the Tuesday made a false and self-serving report to police that he had seen a fire fighter driving a four wheel utility lighting one of the fires. Justice Coghlan characterized this report as "a deliberate and careful attempt to attach the blame to others." (at [25]) Later Sokaluk acknowledged to police that he had made the report so that he would not be blamed for the fire. When police spoke to him some days later, he told them that he had been smoking in his car and asserted that a piece of paper must have ignited, after which he panicked, and some time later reported the fire to authorities. However, Coghlan J noted that expert evidence repudiated the feasibility of his account. How these situations were interpreted by the sentencing judge was that since Sokaluk was able to function to some extent in an apparently reasoned and sophisticated way in the aftermath of his fire-lighting he was significantly culpable for his criminal conduct because of having the capacity to understand the nature of what he was doing. The legitimacy of such judicial reasoning is a function of whether the two scenarios were properly commensurate for a person with ASD. At the time of writing, the case is subject to appeal.

The sentencing judge reviewed in some detail the catastrophic nature of the fire and drew particular attention to the "self-sacrifice and courage" of the volunteers in the area who fought the fire. He took into account the hurtful nature of the way in which the fires had started because of Sokaluk and the life-changing nature of the fires for those who survived them.

Justice Coghlan received a substantial amount of material about Sokaluk who was 42 years of age at the time of sentencing and had no prior convictions. He received multiple expert reports which led him to conclude that Sokaluk suffered from an ASD and was intellectually disabled to a "reasonably mild degree" (at [46]). He noted that Sokaluk had grown up in the local area and had experienced difficulties during his schooling. He attended a "special school" for children with disabilities but managed to gain employment at a university where he worked as an assistant gardener for some 16 to 17 years. Justice Coghlan concluded that Sokaluk was teased and perhaps bullied in the workplace as he had been at school. Interestingly, and perhaps suggestive of his having a problematic interest in fires, he also worked for the Country Fire Authority but ceased employment in about 2006 and went onto a disability pension. He owned his own house and lived there alone but was dependent on his parents for cooking, cleaning and managing his finances. He had had two serious relationships with women and was closely emotionally connected to his dog.

Justice Coghlan accepted that Sokaluk had a "mental impairment" for the purposes of sentencing by reason of his conditions of autism and intellectual disability. He had regard to what he classified as Sokaluk's "reduced moral culpability" and stated that he had moderated general deterrence as a factor to which he had regard in sentencing. He concluded that "personal deterrence looms somewhat larger for you than it might for others." (at [66]) He stated that he regarded Sokaluk as "genuinely remorseful" and accepted that he did not "set out to achieve this awful result" (at [66]). He accepted that the sentence of imprisonment that that he imposed would "weigh more heavily upon you than on others" (at [66]).

So far as expert evidence was concerned, Coghlan J received multiple forensic mental health expert reports but had particular regard to that of Professor James Ogloff, the Director of the Centre for Forensic Behavioural Science and Foundation Professor of Clinical Forensic Psychology at Monash University. Professor Ogloff observed that Mr Sokaluk:

> "would occasionally stare blankly ahead. ... He did not appear emotionally distressed or anxious. He displayed repetitive motor behaviour which consisted of lightly touching the edge of the table that separated us and moving his hands together and apart slowly. This behaviour subsided over the course of the interview. Mr Sokaluk demonstrated very concrete literal thinking. He appeared emotionally blunted and socially immature. (Psychological Court Report, 22 December 2012, para 5)

Professor Ogloff noted Sokaluk's assessed level of intellectual functioning, measured overall at an IQ of 74 and another psychologist's assessment that his profile was typical of a person with autism. He took into account that Sokaluk's relative strengths were in the areas of visual perception, non-verbal processing and attention to visual detail, while his weaknesses were in the ability to comprehend and/or respond to questions. Professor Ogloff expressed the view that the autistic symptoms experienced by Sokaluk "have been debilitating and dysfunctional, resulting in difficulties in relationships, employment and general life skills." (Psychological Court Report, 22 December 2012, para 9). He emphasized Sokaluk's response to the question of who the person was who was most important to him. Sokaluk responded that it was his dog and related very detailed, anthropomorphizing accounts of his dog. While Sokaluk was dependant upon his father, there was little evidence of emotional connection with him.

Professor Ogloff concluded that Sokaluk was fit to stand trial, although he thought he would experience some difficulties in following the evidence, and that the defence of mental impairment (insanity) was not available to him. Professor Ogloff expressed the view that Sokaluk was not a pyromaniac but did not feel able to identify with confidence the characteristics or motivations which had led him to engage in his fire-setting behaviours, other than to say that if he did deliberately light the relevant fire his motivation was probably expressive (namely a means of emotional expression, given his social inadequacies)

http://dx.doi.org/10.5772/55400

Mr Sokaluk meets the criteria for a diagnosis of Autism Spectrum Disorder. This disorder has affected his social and adaptive functioning all of his life. He does not meet the criteria for a diagnosis of a major mental illness or personality disorder at present, although he has been treated with medication in the community for depression and in prison for lowered mood and anxiety.

Whilst his overall level of intellectual functioning is in the borderline range, his verbal capacity is more limited and, in fact, falls in the intellectually disabled range. Conversely, his perceptual capabilities are much better, falling in the low average range. This suggests that while Mr Sokaluk has been able to hold a job, operate a motor vehicle, and live on his own, his level of intellectual reasoning and verbal comprehension is very impoverished. He has been dependent on his parents for maintaining his finances, cleaning his house, and providing him with meals. It takes him much longer to acquire information or to learn a task than would be the case for most others and his abstract reasoning capacity is very limited. His presentation, reasoning, receptive and expressive language are affected by the confluence of his Autism Spectrum Disorder and decreased level of intellectual functioning. For example, he is a very concrete and literal thinker.

Justice Coghlan sentenced Sokaluk to a total effective sentence of imprisonment of 17 years and 9 months, with 14 years to elapse before he would be eligible for parole. Both the Director of Public Prosecutions and Sokaluk appealed against the sentence, the one contending it was too short, the other that it was too long. At the time of writing the appeal had not been heard.

The Sokaluk appeal raises for consideration the relevance of ASD (and an intellectual disability) to the evaluation of criminal culpability when the defendant's capacity to appreciate the consequences of his behaviour is reduced by reason of his disorder. Expert evidence before the court suggested that Sokaluk's intellectual and abstract reasoning were at a low level, that his emotional connections with people were poor and that he had been ill-treated over a period of time by reason of his difficulties in communicating and interacting with others. However, Sokaluk was far from wholly disabled and to varying extents had been able to function within the community and had some capacity to appreciate that forms of behaviour are unacceptable and wrong. Thus, the question arises as to how severely he should have been punished and deterred from conduct whose terrible repercussions he was found not to have set out to achieve and whose ability to foresee and appreciate was unclear.

7. State of Western Australia v Mack (2012)

Brent Mack was charged with the murder of his mother but in an application that he made for a judge-alone trial questions arose about his fitness for trial on the basis of his suffering from autism. His counsel swore an affidavit expressing the view that there was a risk that Mack would not participate in the trial in any way, including the provision of instructions, thus making his defence extremely difficult.

A psychiatrist retained for Mack expressed the view that Mack was unfit to stand trial because of his inability to follow the course of the trial and to defend the charges against him:

He has impairment in the use of multiple nonverbal behaviours, including eye contact and body posture; a lack of social reciprocity; the failure to develop any appropriate peer relationships. He also exhibits impairments in communication in relation to the inability to sustain a conversation; stereotyped use of language; monotonous speech with an abnormal, robotic rhythm to it; and inability to understand the nonliteral aspects of communication or applied meaning.... [H]is ability to understand the abstract is virtually absent and everything is very much concrete interpretations of things.(*State of Western Australia v Mack*, 2012: [19])

When asked about the contrast between this presentation and Mack's manner in his records of interview, the psychiatrist stated that he had heard Mack speaking in a similar way to his responses in the records of interview when he spoke to members of a working party about native plants, a subject in which he had a particular interest. He expressed the view that Mack did not process emotion, particularly negative emotion, his response tending to be one of retreat from a situation physically or into himself. In such circumstances his deficits in short term memory and concrete thinking were exacerbated. He expressed the view that:

Mr Mack tends to be quite dichotomous in his thinking, so from my assessment of him, he divides things into a personal context - that's his language - personal context or some other context, such as a business context. If anything is relevant to himself personally, he tends to have a somewhat all-or-nothing approach to that. So he obliterates that from discussion completely. If it's something to do with something about which he's factually knowledgeable, then he's probably happy to talk at length about it. ... from my interviews with Mr Mack, I would expect that he would be very reluctant to talk about any matters that might arise during the course of the trial (*State of Western Australia v Mack*, 2012: [24]).

A psychiatrist called by the prosecution, although agreeing on the diagnosis, and conceding a potential for Mack's cognitive ability or performance to deteriorate during the trial because of anxiety, expressed the view that he was fit for trial. He accepted that Mack had a propensity to focus on the way in which questions were asked, rather than their substance but concluded there was no evidence to "suggest the presence of any difficulties in registration" of the content of communications to which he was privy.

Justice McKechnie accepted that the behaviour of Mack was unusual but found the evidence of the psychiatrist called by the prosecution to be more consistent with the performance of Mack during his records of interview. He concluded that "It is likely that the accused's current presentation is more as a result of choice coupled with his autism than a result simply of his mental impairment." (*State of Western Australia v Mack,* 2012: [24]). He placed little weight on the submission that Mack's odd presentation might cause him prejudice before jurors who might be distracted by it or draw adverse inferences against him (cp *McGraddie v McGraddie,* 2009; *Parish v DPP,* 2007). However, he concluded that because of Mack's autism and its impact on the trial process generally, the interests of justice weighed in favour of a trial by judge alone (*State of Western Australia v Mack,* 2012: [44]).

The decision of McKechnie J highlights the difficulty encountered by those with ASD in being able effectively to communicate with and give instructions to their lawyers in the unwonted and intimidating atmosphere of the courtroom. While persons with ASD may be articulate and contextually appropriate when conversing about a subject of interest or fascination that is non-threatening, a wrong inference may be extrapolated that they are capable "if they simply make an effort" of speaking with their lawyers, understanding testimony and its import for their defence, and giving evidence in a courtroom. While the discontinuity between these contexts is not immediately obvious, the nature of ASD, if well explained by a mental health expert, has the potential to be compelling.

Another aspect of Mack's case that is significant is the failure of the trial judge to accept that the conduct of the defendant might be highly prejudicial and, in particular, be misinterpreted and misconstrued by a jury. This is a problematic issue for defendants with ASD because of their propensity to conduct themselves oddly and with apparent disinterest in the circumstances of their victim and the ramifications of their conduct. There is often a risk that their manner, their words and their reactions may lead jurors and judicial officers who are uninstructed in the characteristics and symptomatology of ASD to draw wrong and damaging inferences (see eg *McGraddie v McGraddie,* 2009; *Parish v DPP,* 2007).

8. Challenges for mental health expert evidence

The 2011/2012 decisions by courts in *HPW, Sokaluk* and *Hampson* illustrate the risk that ASD will not to be found by judicial officers to have a major relevance for the determination of criminal culpability. What each case has in common is conduct that is such as to prompt high levels of censure by reference to ordinary community standards and thus a risk that such considerations will overbear subtle issues relating to the personal blameworthiness of an offender. However, there is reason to suspect that in each case the defendant's ASD constituted at least a significant context within which the criminal conduct was committed and there is reason to postulate that it may have had a sufficient influence on the conduct to have been a genuinely mitigating factor in terms of each offender's moral blameworthiness.

The decisions of *George, Sokaluk* and *Mack* are exemplary of cases where wrong inferences may be drawn by reason of the capacity of persons with ASD to conduct themselves in ways comparable to how others with full capacity might behave. There is a need in many criminal trial contexts that deal with persons with ASD for expert evidence that is counter-intuitive and directed toward the need for care to be taken by decision-makers, judicial or lay (ie jurors), in drawing inferences on the basis of otherwise known conduct and capacities of persons with ASD, especially when different scenarios in defendants' lives are compared. Capacity is highly situation-related, and for all of us is variable by reference to context. Persons with ASD may be high-functioning in some contexts but when comfort zones are intruded upon or when they are outside an environment that is structured or familiar, their conduct may be erratic, their judgment poor and their capacity to appreciate the resonances and repercussions of their actions limited. This is relevant both to their capacity to function effectively within a trial context, including their fitness to stand trial, and to their criminal responsibility and moral blameworthiness for actions for which they are being tried.

Fitzgerald (2010) has postulated a subcategory of ASD that he calls "Criminal Autistic Psychopathy", characterised, he says, by persons with callous, unemotional traits who repeatedly engage in anti-social criminal conduct. He has instanced a number of serial killers who he maintains have combined features of ASD and Psychopathy. While there are theoretically fundamental differences between the two disorders, the former for instance being a developmental disorder, Fitzgerald makes a persuasive argument for the overlap of traits/ symptoms in some persons. For these individuals the existence of conjoint pathology or a hybrid disorder is particularly problematic at sentencing as it is most likely to arouse concerns in relation to the need for protection of the community rather than an empathic focus on impaired levels of moral blameworthiness.

Finally, two other important issues consistently arise in criminal cases. The behaviour of a person with ASD at trial can be alienating and highly prejudicial. This bears upon whether they should be accorded the opportunity for a judge-alone trial, where that facility exists (see eg *State of Western Australia v Mack*, 2012), or whether expert evidence to disabuse jurors of misimpressions they might otherwise form should be permitted. In addition, the capacity of a person with ASD to cope without decompensating, being dangerously victimised or having the anxiety and depressive symptomatology, which is often part of an ASD (see eg *R v Sokaluk*, 2012), exacerbated within a custodial environment frequently needs to be the subject of expert opinion evidence from professionals with a sound understanding of the impact of ASD on day-to-day functioning for those with the disorder.

The challenge for mental health professionals who seek to educate courts about the relevance of ASD to decision-making about accused persons' responsibility for criminal conduct and their blameworthiness for their actions lies in identifying the causative role of ASD and its repercussions for the imposition of custodial sanctions. The reality of ASD is that it is easily misdiagnosed as it can easily fail to be identified, it can it can co-exist with a variety of other disorders – anxiety, depressive and personality - and it can be highly exculpatory or at least explanatory. On other occasions though it is no more than part of a context and is not particularly mitigating at all. More than simply identifying the disorder by correct diagnostics, the real issue for mental health professionals is to evaluate in a rigorous and informed way how ASD fits into the picture of criminal culpability for a particular individual in respect of particular conduct at a particular time.

http://dx.doi.org/10.5772/55400

9. Cases

IA v The Queen [2005] EWCA Crim 2077.

McC v The Queen [2007] NSWCCA 25.

McGraddie v McGraddie [2009] ScotCS CSOH 142.

Parish v DPP (2007) 17 VR 412; [2007] VSC 494.

R v George [2004] NSWCCA 247.

R v Hampson [2011] QCA 132.

R v Mueller (2005) 62 NSWLR 476; [2005] NSWCCA 47.

R v Sokaluk [2012] VSC 167.

R v Somogyi, 2011 ONSC 483.

State of Western Australia v Mack [2012] WASC 127.

ZH v The Commissioner of Police for the Metropolis [2012] EWHC 604.

Author details

Ian Freckelton[1,2]

1 Melbourne Bar, Victoria, Australia

2 Law Faculty, Department of Forensic Medicine and School of Psychology and Psychiatry, Monash University, Australia

References

[1] American Psychiatric Association. (2011). DSM-V: Autism Spectrum Disorder. http://www.dsm5.org/proposedrevision/pages/proposedrevision.aspx?rid=94,viewed 12 October 2012.

[2] Barry-Walsh, J, & Mullen, P. (2004). Forensic Aspects of Asperger's Syndrome. *Journal of Forensic Psychiatry and Psychology 15*(1), 96-107.

[3] Browning, A, & Caulfield, L. (2011). The Prevalence and Treatment of People with Asperger's Syndrome in the Criminal Justice System. *Criminology and Criminal Justice.* , 11, 165-180.

[4] Cascio, C. J, Moana-filho, E. J, Guest, S, Nebel, M. B, Weisner, J, Baranek, G. T, & Essick, G. K. (2012). Perceptual and Neural Response to Affective Tactile Texture Stimulation in Adults with Autism Spectrum Disorders. *Autism Research* (epub).

[5] Cashin, A, & Newman, C. (2009). Autism in the Criminal Justice Detention System: A Review of the Literature. *Journal of Forensic Nursing.*, 5(2), 70-75.

[6] Debbaudt, D. (2002). *Autism, Advocates and Law Enforcement Professionals.* JKP: London.

[7] Fitzgerald, M. (2010). *Young, Violent and Dangerous to Know.* Nova Publications: New York.

[8] Freckelton, I, & List, D. (2009). Asperger's Disorder, Criminal Responsibility and Criminal Culpability. Psychiatry, Psychology and Law , 16(1), 16-40.

[9] Freckelton, I. (2011). Asperger's Disorder and the Criminal Law. *Journal of Law and Medicine* , 18, 677-694.

[10] Freckelton, I. (2011). Autism Spectrum Disorders and the Criminal Law. In M-R Mohammadi. A Comprehensive Book on Autism Spectrum Disorders. Intech: Croatia. , 249-272.

[11] Freckelton, I. (2012). Hoarding and the Law. *Journal of Law and Medicine.* , 20, 225-249.

[12] Freckelton, I. (2013). Autism Spectrum Disorders and the Law. *Journal of Applied Research into Intellectual Disability.* (in press).

[13] Freckelton, I, & Selby, H. (2013). *Expert Evidence: Law, Practice, Procedure and Advocacy.* Thomson-Reuters, Sydney.

[14] Haskins, B. G, & Silva, J. A. (2006). Asperger's Disorder and Criminal Behavior: Forensic Psychiatric Considerations. *Journal of the American Academy of Psychiatry and Law. 34*, 374-384.

[15] Kristiannsson, M, & Sorman, K. (2008). Autism Spectrum Disorders: Legal and Forensic Psychiatric Aspects and Reflections. *Clinical Neuropsychiatry* , 5(1), 55-61.

[16] Lane, A. E, Young, R. L, Baker, A. E, & Angley, M. T. (2010). Sensory Processing Subtypes in Autism: Association with Adaptive Behavior. J Autism Dev Disord. , 40(1), 112-22.

[17] Langstrom, N, Grann, M, Ruchkin, V, Sjostedt, G, & Fazel, S. (2009). Risk Factors for Violent Offending in Autism Spectrum Disorder: A National Study of Hospitalized Individuals. *Journal of Interpersonal Violence.* , 24(8), 1358-1370.

[18] Mahoney, M. Asperger's Syndrome and the Criminal Law: The Special Case of Child Pornography: http://www.harringtonmahoney.com/documents/Aspergers%20Syndrome%20and%20the%20Criminal%20Law%20pdf,viewed 10 October (2012).

[19] Mawson, D, Grounds, A, & Tantam, D. (1985). Violence and Asperger's Syndrome: A Case Study. *British Journal of Psychiatry, 147566569*

[20] Mayes T.A. 2003. Persons with Autism and Criminal Justice Core Concepts and Leading Cases. *Journal of Positive Behavior Interventions*. 5(2): 92-100.

[21] Milton, J, Duggan, C, Latham, A, Egan, V, & Tantam, D. (2002). Case History of Co-morbid Asperger's Syndrome and Paraphilic Behaviour. *Medical Science and Law, 42237244*

[22] Mouridsen, S. E, Rich, B, Isager, T, & Nedergaard, N. J. (2008). Pervasive Development Disorders and Criminal Behavior: A Case Control Study. *International Journal of Offender Therapy and Comparative Criminology.*, 52(2), 196-205.

[23] Murrie, D. C, Warren, J, Kristiansson, M, & Dietz, P. E. (2002). Asperger's Syndrome in Forensic Settings. *International Journal of Forensic Mental Health. 1*(1), 59-70.

[24] Realmuto, G. M, & Ruble, L. A. (1999). Sexual Behaviors in Autism: Problems of Definition and Management. *Journal of Autism and Developmental Disorders. 29*, 121-127.

[25] Scragg, P, & Shah, A. (1994). Prevalence of Asperger's Syndrome in a Secure Hospital. *British Journal of Psychiatry.*, 165(5), 679-682.

[26] Silva, J. A, Ferrari, M. M, & Leong, G. B. (2003). Asperger's Disorder and the Origins of the Unabomber. *American Journal of Forensic Psychiatry. 24*, 5-8.

[27] Warren, A. (2006). Asperger's Syndrome and Autism Spectrum Disorders in the Courts. Paper presented at the National Judicial College of Australia Conference: Science, Experts and the Courts: http://njca.anu.edu.au/Professional%20Development/programs%20by%20year/2006/Science%20and%20courts/Anthony%20Warren.pdf,viewed 10 October 2012.

Section 6

Architecture and Autism

Chapter 9

Autism and Architecture

Francisco Segado Vázquez and
Alejandra Segado Torres

Additional information is available at the end of the chapter

http://dx.doi.org/10.5772/53679

1. Introduction

"At the International Congress "Building, Dwelling, Thinking" held in 2001, Heiddeger concludes by highlighting the convenience and importance for the scientific architect to develop architecture by *"building from living and thinking about dwelling"*.

Architecture has been defined in many ways throughout history, but its focus, its aim, its purpose, is dwelling. For this reason, Norberg-Schulz (1980) affirms that in order to research and understand an architectural space, it is necessary to understand *existential* space, that is, the concept of space that allows man to create a stable image of what surrounds him, at the same time allowing him to belong to a society and culture.

In an architect's work, there is an underlying notion, which may be evident to a greater or lesser extent, that the built environment is a space that is to be *lived in*, inhabited, for it to be considered architecture. It is this existential experience of the space which gives it a sense of place and not a mere sense of the abstract.

Likewise, for many years, architecture has taken into account the existence of people with different types and degrees of disabilities (mainly visual, hearing and motor), and the architect has planned and designed, either in accordance with their convictions or purely down to legal guidelines, so that spaces can also be inhabited by these people. So, here we are talking about "accessibility", which is a clearly (although not exclusively) physical concept: this is a matter of enabling disabled people to access buildings/spaces, which subsequently makes it possible for them to inhabit them.

However, there are other deficiencies or disabilities that are not so "visible", and that are obviated in making a built environment "accessible". According to Dianne Smith (2009), in the design process (of a building, of a street, of a town, of an interior space...) two paradigms intervene, almost exclusively: that of the client/property developer and that of the architect. That is, it is the visions that these two agents have of reality, of how things work and are perceived, which give shape to the building. This, moreover, on numerous occasions, with the prior assumption that said environment is to be practically limited to being a container or backdrop for certain activities or functions.

Nonetheless, for people with certain cognitive and sensory deficiencies, etc., which are *"less visible"*, as Smith herself affirms, including people suffering from autism, this supposition regarding how spaces are to be perceived and inhabited is far from the truth: due to their deficits, they have to make an effort, sometimes an enormous one, to be able to assimilate and understand the environment surrounding them. In this struggle, due to the problems that they have in processing the information that they receive via their senses, many factors may imply a great barrier and, at certain times, may cause a "blockage" in their comprehension of the environment, which, at the same time, leads to frustration and strange behaviour in the eyes of a chance observer (gestures, verbal expression, movement…).

Therefore, the surroundings, the built environment, is a factor which notably affects (directly and in many other indirect ways) people with certain *less visible* deficiencies. As the architect John Jenkins states, with reference to the design of educational areas for autistic children, although it may be generalised to people of any age and to other types of buildings, *"mainstream children are probably more 'able to cope' with badly designed spaces than an autistic child would be. So the responsibility to create a 'good' environment is brought into sharp relief."* (Quoted in Humphreys, 2008, pg.41).

2. Autism

In this section, the intention is to give a global vision of what is understood by the term 'autism', and what the characteristics of people with autism are.[1] It is true that the definitions of the disorder, its etiological explanations, the nosological considerations, and even the treatment of people with autism, have changed over time, in keeping with the progress that has been made in research into autism from diverse, although complementary, fields such as medicine, psychology, pedagogy or even philosophy. However, it is necessary to know what peculiarities people with autism show in order to determine what the characteristics are that a built environment has to have in order to make it easier for them to grasp and so achieve other objectives that go beyond, but to a certain point depend on, the architecture itself, such as encouraging learning, promoting autonomy, making it easier to socialise, ensuring independence or even preserving the dignity of the person with autism.

1 The intention is not to assert that the characteristics are unique. Each person with autism shows symptoms in an almost unique manner. It is a matter of seeing the common characteristics, aspects and behaviour that are frequently apparent in people with autism.

http://dx.doi.org/10.5772/53679

2.1. General concept

Autism is one of the most fascinating disorders that medicine and psychology have had to face. Isolation or solitude is one of the most enigmatic characteristics of autism. In fact, when American psychiatrist Leo Kanner (1943) describes the autistic disorder for the first time, he points out that the pathognomic sign is the inability to relate to other people, which causes an *"extreme autistic solitude"*. In this first description, Kanner specifies a series of common characteristics in the children that he studied, which we can summarise below:

- Inability to relate to other people, at least in a normal way
- Extreme autistic solitude which apparently isolates the child from the outside world
- Deficiencies in the language, which may include muteness, pronominal inversion, echolalia or an idiosyncratic way of speaking, among others
- In some cases, an excellent literal memory
- Preference for certain specific foods (from a very young age)
- Fear of intense noises
- An obsessive desire to repeat and insistence on an invariable environment[2]
- Scarce repertoire of spontaneous activities (like normal play)
- Strange motor stereotypes, like spinning or swaying
- Normal physical appearance
- Appearance of the disorder in the first three years of life

2.2. Historical evolution of the consideration of autism

During the years prior to the appearance of Kanner and Asperger's articles, as a consequence of the wide diffusion of psychoanalytical theories, and in spite of the fact that Kanner himself had suggested a biological deficiency, it was considered that autistic disorder had a psychodynamic aetiology, that is, that it had originated due to emotional causes, leading to the blame being laid on the parents (there was talk of cold mothers, unaffectionate fathers…). So, it was finally affirmed that the cause of autism was the parents' wish for the child not to exist (Bettelheim, 2001). The psychoanalytical therapies used tried to restore these alleged emotional wounds and reconstruct the supposedly broken affections. This type of psychodynamic treatment, in the opinion of many contemporary researchers, has not made many contributions. (for example JK Wing, 1968: Escobar Solano, Caravaca Cantabella, Herrerro Navarro and Verdejo Bolonio, s.d.).

2 The term used by Kanner is sameness, which could be interpreted as "similarity" or "monotony", but none of these two words can completely describe the original meaning (situation in which there are no changes). This is often interpreted as "invariance in the environment" or "Kanner's autism"

From the mid 1960s until around the middle of the 1980s, autism has gone from being considered an emotional disorder to the opinion that it has a neurological origin, finally being treated as a cognitive disruption, rather than affective (Escobar Solano et al., s.d.). Methodical and rigorous research began on autism, to try to understand alterations in communication and language, as well as in social relationships, resistance to change, etc (for example, Rutter and Schopler, 1984; L. Wing and Gould, 1979).

From that moment, and thanks to the progress made in research, autism is now considered to be a developmental disorder. Autism is included among the so-called Generalised Development Disorders, which, as well as autistic disorder[3], include others such as Asperger's Syndrome, Rett's Syndrome, child disintegrative disorder, and the non-specified generalised development disorder. Recently, it has also come to be understood that on many occasions it is not easy to set a clear limit among these disorders, instead there is a type of continuum in which three essential areas are affected to a greater or lesser extent[4]: communication (verbal and non-verbal, as this does not only refer to language), social reciprocity, and the absence of imaginative behaviour and symbolic play, with highly repetitive interests and activities. For this reason, talk of Autistic Spectrum Disorders (ASD) came about, which nowadays is a common term (in fact, the upcoming APA Diagnostic and Statistical Manual of Mental Disorders, DSM-5, which is hoped to be published in 2013, considers this denomination).

3. Design criteria

We will go on to present, fleetingly and not in great depth, some aspects of people with ASD to whom a solution can be given using architectural project and design mechanisms. We will group them, in order to make their presentation more systematic, according to the different areas that may be affected in said people.

3.1. Imagination

Resistance to change and a limited capacity of imagination are one of the essential characteristics of autistic spectrum, and these are reflected in aspects such as difficulty or extreme nervousness when changing activity, and even when moving from one space to another (because people with ASD are incapable of "imagining", in the sense of creating a mental image of what there might be at the other side of a door or wall, for example). From an educational point of view (and even in family life) this aspect is faced by "anticipating" the activities that are going to be carried out next, and avoiding or lessening, as far as possible, unexpected changes in the planned routines.

From the point of view of an architectural project, the inability to construct a mental image of the environment, as well as to integrate parts into a whole, may be faced by

3 Which would correspond with the so-called "classic autism" or "Kanner's autism"

4 This is known as "Wing's triad" (L. Wing and Gould, 1979)

http://dx.doi.org/10.5772/53679

looking for a clear structure in the building, as well as by providing elements that give it a certain order and unity, in such a way that the building can be easily read, predictable, *imaginable*. Referring to the transition between spaces, the anxiety suffered by people with ASD can be reduced for example, by using colours on the doors (depending on the spaces behind them), as well as pictographs and photographs which "advance" what we are going to find, or by creating transitional environments in between, where the change of space can be anticipated.

3.2. Communication

Difficulties in verbal and non-verbal communication, together with difficulties processing information, make it necessary to "remove certain psychological "barriers" and adapt the environment with codes which [...] are characterised by being specific and easily perceivable (as opposed to subtle), simple, that is, containing few elements (as opposed to complex) and permanent (as opposed to temporary)" (Tamarit, De Dios, Domínguez, and Escribano, 1990).

The person with ASD needs visual support for communicating and pictographs or photographs of objects, people, etc. are usually used. The built environment should be able to "welcome" these forms of communication, foreseeing their correct location and integration. Colour coding, for example, of different elements may also help to improve communication.

3.3. Social interaction

Difficulties in social interaction are taken into account, by definition, although to varying degrees, in people with ASD. For this reason, different educational strategies try to influence this aspect. Therefore, it will be necessary to provide the spaces in which to allow and even encourage social interaction, although always taking into account that people with ASD may show particular proxemics[5], needing wide, open spaces, in which said interaction may take place without getting too close. A combination of larger areas and others, in which interaction can take place more closely, if required, would be advisable. Moreover, at certain times a person with ASD may feel overwhelmed by a demanding social situation (in the sense that they are forced to participate in several interactions) and need a space to which they can retreat in search of privacy or a "simpler" interaction (less people, or people with whom they are more familiar).

3.4. Sensory difficulties

In the case of ASD sufferers, it is also common for malfunctions to exist in the reception (or the processing) of stimuli, which is demonstrated by a visual, acoustic, vestibular or tactile (although also often related to smell or taste) hypersensitivity (or sometimes hyposensitivity). The proprioceptive sense is also altered at times. A consideration of this aspect should lead us to be careful when designing with colours (which do not clash excessively, are not

5 Proxemics refers to the space that exists between people in different social interactions. Proximity may be perceived as a threat by a person with autism.

too strong or too bright), textures or patterns, with acoustic properties in these spaces and the construction elements separating one from another, with lighting (soft lighting is recommended, preferably sunlight, and in all cases avoiding fluorescent strip lights, as the flickering and buzzing may upset a person with visual or auditory hypersensitivity),as well as with the fittings, etc. Another example of sensory alteration is a different perception of the sensation of pain, which may mean that a person with ASD could suffer serious burns on their hands, due to not moving them in time when water from a tap, for example, comes out at a very high temperature, or they may have a serious cut or injury and barely notice.

Multi-sensory stimulation rooms ("Snoezelen" rooms) allow people with ASD to adjust their sensory perceptions and also reduce anxiety at specific moments.

3.5. Behaviour and safety

Behavioural problems are also frequent in people with ASD, and may lead to aggressive conduct, meaning that the elements in the built environment have to be designed, chosen and implemented taking into account these potential bouts of aggression. Examples of these elements that are to be paid special attention to are bathroom fittings, electrical devices, metal door fittings, banisters and railings, exterior carpentry, tiles, etc.

4. Conclusion

It has been proven that existing scientific literature regarding built environments in relation to people with ASD and vice versa is scarce, and this is in spite of significant research activity carried out in relation with autism in recent years. This interest is due to the significant increase in the number of cases diagnosed, meaning that prevalence studies produce much greater ratios than the figures of 1 to 3 people in every 10,000 that were handled at the beginning of the 1990s and which were previously even lower. Recently it has been affirmed that there is one child with ASD in every 110 born (CDC – Center for Disease Control and Prevention, 2009). It is clear that the increase in numbers does not reflect (at least not exclusively) a real increase in the number of cases, but the expansion that the concept of autism has undergone, stretching to that of autistic spectrum, and to health care and education which allow for early diagnosis, with a greater awareness of the existence of the disorder (Ahrentzen and Steele, 2009). In spite of this, figures reveal that it is a significant group of the population, which requires attention from society. In our area of discipline this should also be the case. In just a few years, architects have been made aware of how to draw up plans without the so-called "architectural barriers" that limit accessibility for people with a disability. However, under this concept of a barrier, we do not usually include those which limit the use of the built environment for people with cognitive or mental disabilities. As Baumers and Heylighten (2009, 2010) state, these people perceive space in a unique, different way, with the "mind's eye".

It is necessary to progress in research in this sense, analysing the architectural achievements designed and built for people with ASD, checking how suitable they are for the particular characteristics of this part of the population, even studying any defects they may have and verifying the new contributions that can be made in them.

It would also be interesting to encourage field studies with specific interventions in the built environment, even on a smaller scale, such as that of Magda Mostafa (2008), which allows us to extract results that can be checked and verified on how certain activities improve, and to what extent, the experience of the person with ASD in their built environment.

If, in general, the constant reflection upon the relationship between the person and space, between the individual and their environment (built), is important for the discipline of architecture, we believe that the particularisation of this reflection for the *dweller* with autism may be an interesting contribution for the discipline itself. In fact, researching about this adjustment and this link, between the architectural object and its aim - the person, is to reflect upon architecture itself, which, like other arts and other disciplines such as Philosophy, grows upon *rethinking*.

Finally, we will conclude with a quote from Luis Fernández-Galiano, which allows us to situate the role of the architect, especially in the case of people who are to be found "within the spectrum":

"Dwelling is a difficult job. Like the profession of living, that of dwelling requires continual learning and attention, demands meticulous, systematic effort, and claims an immeasurable investment of time and energy. The nature with which the majority of people manage to carry out the complicated rituals of the dwelling space is surprising. Just as happens in the case of language, expertise in use is acquired along with habit, which provides guidelines and domesticates gestures and voices via daily reiteration of movement and words. So, this tiring and habitual profession has both an obstacle and an accomplice in the architect" (quoted in Oyarzun, 2005)

Author details

Francisco Segado Vázquez[1] and Alejandra Segado Torres[2]

1 Polytechnic University of Cartagena, Spain

2 Faculty of Medicine – Complutense University Madrid, Spain

References

[1] Ahrentzen, S., and Steele, K.(2009). Advancing full spectrum housing. Phoenix, USA: Arizona Board of Regents.

[2] American Psychiatry Association. (2002). DSM-IV-TR, Diagnostic and Statistical Manual of Mental Disorders, Masson, S.A.

[3] Asperger, H. (1991). Autistic psychopathy in childhood. In U. Frith (Ed.), U. Frith (Trans.). Autism and Asperger syndrome (pg. 37). Cambridge: Cambridge University Press.

[4] Balbuena Rivera, F. (2007) Breve revision histórica del autismo (A brief historical review of autism). Revista de la Asociación Española de Neuropsiquiatría, 27(2), 61-81. (Spanish Neuropsychiatry Association).

[5] Baron-Cohen., and Heylighen, A. (2009). The Eyes of the Mind. Architecture and Mental Disability. In Engaging Artefacts. Presented at the Nordic Design Research Conference – NORDES'09. Oslo.

[6] Baumers, S., and Heylighen, A. (2010). Harnessing Different Dimensions of Space: The Built Environment in Auti-biographies. In P.Langdon, P.J. Clarkson, and P. Robinson (Eds.). Designing Inclusive Interactions: Inclusive Interactions Between People and Products in Their Contexts of Use (pg. 13 – 23).

[7] Beaver, C. (2010). Autism-friendly environments. The autism file, (34), 82-82.

[8] Bettelheim, B. (2001). The Empty Fortress: Infantile Autism and the Birth of the Self. (Spanish version) Saberes Cotidianos. Barcelona: Paidós.

[9] CDC – Center for Disease Control and Prevention. (2009). Prevalence of Autism Spectrum Disorders. Autism and Developmental Disabilities Monitoring Network, United States, 2006. MMWR, Surveillance Summaries, 58(SS-10), 1-20.

[10] Escobar Solano, M., Caravaca Cantabella, M., Herrero Navarro, J., and Verdejo Bolonio, M. (s.d.) Necesidades educativas especiales del alumnado con trastornos del espectro autista. (Special educational needs of the pupil with autistic spectrum disorders) Pending publication.

[11] Frith, U. (2006). Autism: Explaining the Enigma. (Spanish version). Madrid: Alianza Editorial.

[12] Frith, U. and Happe, F. (1999). Theory of Mind and Self-Consciousness: What Is It Like To Be Autistic? Mind and Language, 14(1), 82-89.

[13] Grandin, T. (1992). An inside view of autism. In E.Schopler and G.B. Mesibov (Eds.), High-functioning individuals with autism (pg. 105-126). New York: Plenum Press.

[14] Heidigger, M. (2001). Lectures and articles. (Translation:E. Barjau). Barcelona: Ediciones del Serbal.

[15] Humphreys, S. (2008). Architecture and Autism. Recovered from http://www.auctores.be/auctores_ bestanden/UDDA%2003102008%20S%20Humphreys.pdf

[16] Kanner, L. (1943). Autistic disturbances of affective contact. Nervous child, 2(2), 217-230.

[17] Khare, R., and Mullick, A. (2008). Educational spaces for children with autism; design development process. In CIB W 084 Proceedings, Building Comfortable and Liveable Environment for All (pg. 66-75). Atlanta, USA.

[18] Khare, R., and Mullick, A. (2009). Incorporating the Behavioural Dimension in Designing Inclusive Learning Environment for Autism. International Journal of Architectural Research, 3(3), 45-64.

[19] Lynch, K. (1998). The Image of the City. (Translation by E. Revol). Barcelona: Gustavo Gili.

[20] Mostafa, M. (2008). An Architecture for Autism: Concepts of Design Intervention for the Autistic User. International Journal of Architectural Research, 2(1), 189-211.

[21] Norberg-Schulz, C. (1980). Existence, Space & Architecture. (Translation by A. Margarit Durán) Barcelona: Blume.

[22] Oyarzun, D. (2005). Arquitectura y discapacidad. Centro de atención integral para niños autistas. (Architecture and disability. Integral help centre for autistic children.) (Thesis).Santiago de Chile. Universidad de Chile.

[23] Rutter, M. and Schopler, E. (Eds.). (1984). Autism: A Reappraisal of Concepts and Treatment. (Translation by A. López Lago.) Alhambra Universidad. Madrid: Alhambra.

[24] Scott, I. (2009). Designing learning spaces for children on the autism spectrum. Good Autism Practice, 10(1), 36-51.

[25] Smith, D. (2009). Spatial design as a facilitator for people with less visible impairments. Australasian Medical Journal, 1(13), 220-227.

[26] Southerington, E.A. (2007), Specialized Environments: Perceptual Experience as Generator of Form (Master Project). Cincinnati, USA.

[27] Tamarit, J., De Dios, J., Domínguez, S., and Escribano, L. (1990). Proyecto de Estructuración Ambiental en el aula de Niños Autistas. (Environment building project in the autistic child's classroom). Madrid: Consejería de Educación de la Comunidad Autónoma de Madrid y Dirección General de Renovación Pedagógica del Ministerio de Educación y Ciencia. (Madrid Board of Education and Directorate for Pedagogical Renovation - Ministry for Education and Science.)

[28] Vogel, C.L. (2008). Classroom design for living and learning with autism. Autism Asperger's Digest.

[29] Whitehurst, T. (2007). Evaluation of Features specific to an ASD Designed Living Accommodation. Sunfield Research Institute.

[30] Wing, L. and Gould, J. (1979). Severe impairments of social interaction an associated abnormalities in children: epidemiology and classification. Journal of autism and developmental disorders, 9(1), 11-29.

[31] Wing, L. (1998). The Autistic Spectrum: A Guide for Parents and Professionals. Saberes Cotidianos. Barcelona: Paidós.